Principles of
Environmental Studies

Principles of
Environmental Studies

Aditya Bhatnagar

Oxford Book Company
Jaipur ▪ New Delhi

Published By

Oxford Book Company

Regd. Off. "Bony Residency", Gate No. 2
Opp. Tilak Public School, Vishweswariya Nagar,
Gopalpura Road, Jaipur - 302018, India
Ph: 0141 - 2761280, Telefax: 2761381
email - oxfordbookjpr@gmail.com
website - oxfordbookcompany.com

Delhi Office: 102, 1st Floor, Satyam House,
4327/3, Ansari Road, Darya Ganj,
New Delhi - 110002, India
Ph: 011-45652440

Published in India, 2025

ISBN : 9789380179681

Price : ₹ 2500/-

Typeset by Shivangi Computers, Jaipur

Printed and bound in India by Thomson Press Ltd.,
New Delhi

Preface

Environmental studies is the systematic study of human interaction with their environment. It is a broad interdisciplinary field of study that includes the natural environment, built environment, and the sets of relationships between them. While distinct from ecology and environmental science, the discipline encompasses study in their basic principles as well as the associated subjects, such as: policy, politics, law, economics, social aspects, planning, pollution control, natural resources, and the interactions of human beings and nature.

Principle of Environmental Studies brings to the four various tools, techniques, models and concepts concerning resource management, ecological conservation and wildlife conservation in the perspective of their impact on environment. This book makes elaborate discussion on various connected issues taking specific economic habitats as the base. Agreements, resource management systems, environmental negotiations, landscapes, parks, sanctuaries, wildlife, tourism, and aboriginal life are the topics, many more that are discussed in environmental perspective.

It is written with the understanding that even non-science students should be able to use it. The book Environmental Studies deals with different aspects of environment keeping in view requirements of undergraduate students and people concerned about environmental issues. This will also serve as help book for various competitive examinations.

Aditya Bhatnagar

Contents

Chapter 1

Upper Atmosphere

Some of the atmosphere's profoundest mysteries are far above any possible ascent of man-carrying balloons or propellerdriven airplanes — beyond reach, even, of the twenty-five-mile sounding balloon. In the upper stratosphere there are levels warmed by the action of sun rays on ozone. Beyond the stratosphere, in the ionosphere, there are levels where solar radiation separates innumerable free ions and electrons out of the extremely rarefied air. Some traces of our earth's atmosphere, becoming ever more and more tenuous with increasing distance, probably extend out to cosmic depths of two thousand miles.

Our present knowledge of the upper stratosphere and the ionosphere beyond is mostly indirect — painstakingly gleaned, that is, from such intangible phenomena as sound reflection, light reflection, radio reflection, semi-cosmic clouds, meteors, polar lights, and so on. Or, to put it plainly, our present picture of the upper atmosphere is largely limned with guesswork; and subject to constant correction as Doctor X announces a new and inexplicable observation, as Doctor Y propounds a new and scarcely intelligible theory, or as Doctor Z bobs up to confound Doctor Y. This state of affairs is likely to continue until the sounding rocket at last carries recording instruments, or possibly human observers, above the lower stratosphere into black depths beyond the blue. Nevertheless, innumerable passive observations, as well as certain active experiments with sound waves and radio waves, have already revealed many interesting facts about the upper atmosphere.

ABOVE THE STRATOSPHERE

The colour of the sky, of course, comes from the blue light which the air molecules scatter, and at sunset the light slanting through miles and miles of dense air is predominantly of penetrating red. The duration of twilight is a key, imperfect but not entirely valueless, to the density of the upper air; for the twilight afterglow, so infused with simple and reposeful beauty, is nothing but the shining of air particles high overhead in the rays of the sunken sun. Knowing the time after sunset through which twilight persists, perhaps an hour or two, and the consequent dip of the sun below the horizon, perhaps eighteen degrees or so, scientists have calculated that air sufficiently dense for this reflection extends to a height of about fifty miles.

This fifty-mile altitude is perhaps the upper limit of air as we can perceive it and imagine it — the end of the stratosphere and the beginning of the ionosphere. Also to be seen at dawn and dusk under favorable conditions is the great domed shadow of the earth itself on the upper atmosphere, rising or sinking in that quarter of the sky opposite from the sun a few minutes after sunset or before sunrise. It is seen best in crystal-clear weather from high altitudes. Above the curving shadow-are the sun lights the upper atmosphere with pale rose; below it, there is only dull-gray darkness.* Another sunset phenomenon that tells somewhat of the nature of the upper air is the rose-tinted glow known as the 'purple light,' seen after sunset on very clear evenings above the point where the sun disappeared.

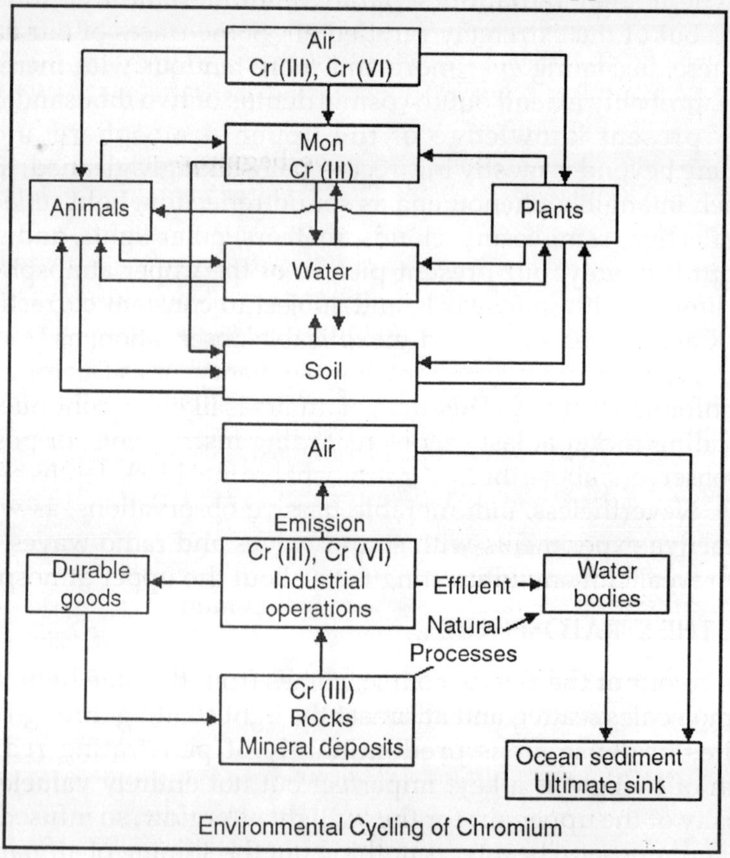

Fig. General Structure of the Air Ocean

It is caused by the refraction and reflection of the longer-wave (red) light through the upper atmospheric layers around the bulge of the earth. An observer out in space (say on the moon during a lunar eclipse) would see it as a continuous ruddy aureole around the whole globe.

Though the stratosphere and the black depths above it are cloudless in the ordinary sense, clouds of a very diaphanous kind are sometimes observed floating at these upper levels. At heights of fifteen to twenty miles patches of parti-colored filmy iridescence, called nacreous or 'mother-of-pearl' clouds, sometimes appear in the crystal-clear skies attending an invasion of cold, dry polar air after the passage of a cyclonic rain area.

These are distinct from, and much higher than, the cirriform clouds that sometimes show iridescence around their edges; in a true nacreous cloud, floating ten miles above the troposphere, the play of colour extends clear into the cloud centre. Nacreous clouds are perhaps composed of microscopic ice crystals or minute, super-cooled water droplets; and they show the existence of mid-stratospheric winds that are mostly from westerly and northwesterly in north temperate and sub-arctic latitudes, ranging in violence from flat calm to more than a hundred and fifty miles an hour.

Considerably higher, marking the beginning of the ionosphere, are noctilucent clouds, occasionally seen after sunset in arctic and sub-arctic latitudes at heights around fifty to fiftyfive miles — heights determined, of course, by triangulation from two or more ground stations situated along known base lines. Noctilucent clouds are excessively thin and wispy in appearance, and their colour shades from golden near the horizon to white and bluish-white above.

Unlike any of the lower clouds, they probably come from outer space and probably consist of excessively fine cosmic dust. Perhaps unable to penetrate the denser atmosphere below fifty miles, they always seem to hang just above this level. Appearing as faintly luminous clouds in the night sky, they are, nevertheless, actually lit by sun rays, and perhaps energized by solar ultra-violet, beyond the earth's shadow. Noctilucent clouds give evidence of superhurricane winds at the base of the ionosphere — drifts from easterly and northerly that sometimes blow at two hundred or three hundred or even six hundred miles an hour!

In 1879, Josef Stefan discovered in old Vienna an important law of physics that might seem at first sight far removed from any usefulness to aerologists. Stefan's law is, in brief, that both the sort and intensity of light rays (both visible and invisible) emitted by any hot body are related to its temperature.

The temperature of the sun is known; therefore, physicists can say how far its radiations extend into the ultra-violet. But it was found that, as we measure them on earth, solar ultra-violet rays stop short of this theoretical spectrum (which extends, with decreasing intensity, even down to X-rays). Hence something in the atmosphere must be absorbing them. That something has turned out to be ozone — the rare triatomic form of familiar oxygen.

Some slight traces of ozone occur near the earth's surface - popularly associated with pine woods and sea breezes. Measurable amounts are formed by electric sparks, and particularly by lightning flashes. But most of the atmosphere's ozone (still only a minute percentage of the other aerial gases) occurs at heights around fifteen to twenty miles — the zone of the nacreous clouds.

At heights of twenty-five to thirty miles the ozone layer is more tenuous, but more important because it receives the full force of solar ultra-violet.

It is a fortunate thing for all God's creatures that this great natural screen of ozone hangs up there thirty miles aloft, incredibly thin and diffused, transparent and invisible, and until recently unknown. For without it, the far ultra-violet and near X-rays shot forth from the sun would destroy all life on the sunlit earth within a few seconds.

But the high-altitude ozone screen has an important effect on the upper air itself. In the words of Haurwitz, 'ozone as a powerful absorber of solar radiation determines to a large extent the temperature of the upper atmosphere.' We have already seen that temperature may increase somewhat with elevation in the lower stratosphere. In the upper stratosphere, where ozone enters powerfully into the picture, temperatures may be much higher — say up to summer heat (around +85° F.) or even near boiling (around +160° F.). These apparent high temperatures have been discovered by nothing more tangible than sound waves.

SKY-REFLECTED SOUND

It is no new fact that powerful noises, such as heavy gunfire, are sometimes distinctly heard at distant places, though entirely unheard at nearer points. Back in 1666, Londoners heard the British Fleet cannonading the Dutch line-of-battle in the Channel, while the sound apparently skipped nearer places along the English coast. This circumstance puzzled Samuel Pepys and other philosophers of that day; and was not explained, in fact, until 1903, by a German named von dem Borne.

Since von dem Borne's time it has often been noticed that the sound of a loud explosion, though heard at great distances, passes completely over intervening points. This peculiar circumstance is best accounted for by supposing that there is in the upper atmosphere some sort of layer which acts in effect as a great sound reflector, and directs outgoing sound waves downward again towards the earth.

In order to fit in with all the other known facts about the upper air, this reflecting 'roof' must take the form of a very warm and very diffuse upper-air layer — it is, in fact, the upper part of the ozone layer. As the velocity of sound is greater in warmer air, the outgoing sound waves travel faster in their upper portions (which encounter high temperatures first), and are hence gradually bent back to earth. The net result is the same as if they had been reflected back from the warm upper layer.

Many experiments to check the existence and height of this sound-reflecting layer were performed in Germany and England soon after the First World War. F. J. W. Whipple, for example, set up some super-sensitive microphones while British artillery was firing on an island in the lower Thames. He heard the gunfire at Birmingham, two hundred and thirty miles away; the sounds had skipped intervening places entirely and seemed to come from the upper air.

Studies of the skip-distance patterns of sound from dynamite explosions, extending out to a distance of six hundred miles, point to the same general conclusion. The only way they can well be accounted for is by assuming high atmospheric temperatures at twenty to thirty miles altitude. The actual reflection summit, at about twenty-five miles, corresponds closely with what is known and assumed about the ozone layer.

The actual pattern of audibility of some great explosion, such as an ammunition dump or a shipload of TNT, is somewhat as follows: The blast is of course heard, and almost felt, out to a distance of perhaps twenty-five or fifty miles. From there to around a hundred miles is a 'skip zone,' or ring of comparative silence. From perhaps a hundred to a hundred and fifty miles extends the first zone of reflected audibility; then a second skip zone; then a second audibility zone from perhaps two hundred and fifty to three hundred miles; also, if the explosion is powerful enough, third and perhaps fourth or fifth skip and audibility zones. These mileage figures are more or less typical, but subject to large variations in practice.

Some recent 'soundings by sound' for upper stratospheric temperatures over California were made by Beno Gutenberg of California Tech. His explosions were the sixteen-inch-gun blasts of the United States Fleet, at gunnery practice off San Clemente Island. His results confirmed the upper temperature determinations made by Whipple and others in Europe. The sound of this battle practice was plainly audible at Santa Barbara, one hundred and twenty miles from the ships. It came in as a dull, jarring, almost soundless rumble that rattled house walls and windows like a mild earthquake.

These upper-air sound reflections are subject to curious vagaries. There are differences between eastward and westward transmission, varying with seasonal effects on the upper-wind distribution. The reflecting layer is warmer and contains more ozone in winter; it is colder and less effective in summer. The curved reflecting layer sometimes serves to intensify the downcoming sound in a peculiar way. In one case, firing with sixteeninch guns actually broke windows in a city nearly two hundred miles away!

The noisiest sound ever heard by civilized men blasted forth when a volcanic island called Krakatoa, between Sumatra and Java, blew up in 1883. The actual blasted-out material, pumice and rock, smashed and covered whole forests on islands miles away. The volcano eruption shot to a height of seventeen miles, and its black pall of concentrated dust extended out one hundred and fifty miles.

An appalling ocean tidal wave rolled out, a hundred feet high, for thousands of miles, swamping hundreds of ships and drowning thirty-six thousand people. The *actual sound* of this prodigious blast was faintly heard at Rodriguez Island, nearly three thousand miles away, and powerfully audible at many places more than a thousand miles distant! The atmospheric wave-disturbance, in sub-audible form, traveled several times around the earth. Also over the whole earth, carried by tropospheric and stratospheric winds during the next few months, spread thin brownish clouds of Krakatoa's dust.

LIGHTS OF THE OUTER AIR

Definite and useful indications about the upper atmosphere are given by meteors, those swift points of light, erroneously called 'shooting stars,' that streak across the starry sky every few minutes on a clear, moonless night. As these bits of cosmic débris bore into the atmosphere at speeds of ten to forty miles per second, it is no wonder that they are intensely heated by friction and soon burst into flame.

If they are observed or photographed simultaneously against a fixed-star background from two or more known locations, their altitude at successive instants can be determined, and it is found that most of them begin to flame about one hundred miles or more above the ground and burn completely out before reaching the thirty-mile level. The larger and slower ones, that is, may continue down to thirty miles, but most of the smaller and faster ones burn out around fifty miles. This knowledge, of course, gives a rough idea of atmospheric density at these heights. Meteors are seen at their best and brightest in the small hours of the morning; for at that time one is on the 'forward' side of the earth as it swings at a comfortable eighteen miles a second along its orbit, and the fiery collisions of the meteors with our outer atmosphere are more or less head-on. Perhaps ten or twenty million meteors enter the earth's atmosphere every day. When we meet a definite swarm of them, this number is vastly increased; during the Leonid swarm of 1833 they appeared as a night-long shower of celestial snow.

Most visible meteors are not much larger than pinheads (ranging, perhaps, from birdshot to buckshot in size) and are completely burned up in the higher atmosphere. But occasionally a larger meteorite, or 'fireball,' weighing several pounds or even several tons and burning brightly enough momentarily to rival the sun, crashes through to the earth's surface. Larger meteors leave trails in the upper air that gleam phosphorescently at night, and resemble lines of white smoke by day.

These trails have been seen to drift rapidly before dissolving, and their motion confirms the idea that enormously high winds can occur at heights above thirty miles. Meteors also confirm another idea about the far upper atmosphere; for the calculated heights at which they should blaze agree with observed heights only if one assumes high temperatures at altitudes around thirty to fifty miles. This result, of course, strengthens the testimony of sound waves and ozone.

Despite their tenuous mystery, introducing new problems, the northern polar lights and their southern counterpart also offer some evidence about the upper air. In northern countries the active aurora is a sight never to be forgotten; widespreading curtains of flame appear and vanish like the draperies of an enormous stage, while arcs and bands of red and green merge into higher rays of violet and blue marching like searchlight beams across the depths of space.

Carl Störmer has observed and photographed the aurora borealis many times in Norway. He finds that it occurs in two general varieties: the lower and more common types, mostly apple-green arcs and bands, flame in the earth's shadow zone at heights around sixty or seventy to one hundred miles.

The higher types, mostly blue, violet and gray-violet rays, flicker at around two hundred miles, or even occasionally as high as four hundred miles, perhaps far up where the sun is shining across the pole above the blackness of arctic night. These heights correspond roughly, as we shall see, with the observed heights of the ionized layers that influence radio transmission.

The spectrum of the aurora shows that the rarefied gases of these far heights are mainly oxygen and nitrogen, rather than the helium and hydrogen formerly thought to predominate in the upper air. There is, in particular, one prominent green line that results when rarefied monatomic oxygen is electrically excited. Moreover, it has been found that this diffuse green glow is always present in the night sky, over all the earth, even above the equator.

It appears to come from atomic oxygen superheated by solar ultraviolet at heights around one hundred to two hundred miles. So the natural human conception of 'seeing your way by starlight' is somewhat of an illusion. Actually about one third of the moonless night sky's luminosity comes from the stars; more than one half comes from this highaltitude oxygen glow.

The primary cause of the aurora is probably 'searchlight' beams of electrons and ions shot out of the sun — beams that rotate with the sun's rotation, somewhat after the fashion of light beams from a rotating beacon, and sweep past the earth. Sometimes, when a brilliant aurora extends visibly to temperate latitudes, this solar 'beam-excitation' can almost be seen.

Thus at midnight the northern aurora may appear from the south as successive outward pulses of light, originating at the left (west) of the sky and flashing progressively towards the right (east) of the sky. These pulses were very apparent in an aurora I witnessed some years ago in New York State. It began with great hanging curtains, but changed in an hour or so to a broad, white band along the northern horizon, surmounted by radial beams of white and pale green.

From observed angles and the known distance of the magnetic pole, this aurora appeared to be about a thousand miles away from me, about a thousand miles wide, and several hundred miles high. The actual speed of the light pulses in the heavens appeared to be around one hundred or two hundred miles a second; and at the earth's distance a rotating solar beam should, in fact, sweep past at about two hundred and fifty miles a second.

Probing the Ionosphere

Within recent years the aurora-haunted depths of the ionosphere have been probed daily and almost hourly by means of upward-sounding radio signals. Radio waves of course speed outward in all directions from the transmitter, along the ground and through the air, at the velocity of light. The ground part of the wave, however, usually dies out within a few score miles, leaving only the 'sky' wave for long-distance transmission and the sky wave, to reach any distant place on earth, must somehow continue to bend around the bulge of our great globe. This bending of the outward-speeding sky wave is accomplished by what amounts to reflection in the ionized layers — layers where solar ultra-violet has disassociated free electrons and ions, increasing the electrical conductivity — of the earth's upper atmosphere.

Actually, just as we saw in the case of the warm ozone layer and sound, an ionized layer does not reflect radio waves; it refracts them by virtue of the fact that the outspeeding electromagnetic vibrations tend to move faster in the more intense ionization of greater heights within the 'reflecting' layer, and hence bend gradually downward again towards the ground. Nevertheless, this refraction process has the same net effect as reflection, so that one may think of the ionized layers as radioreflecting shells or 'roofs' in the upper atmosphere.

Within the last few years it has become possible to measure the height and other characteristics of the ionized layers almost at will. By means of a special transmitter, a radio 'tick,' or signal lasting less than one five-thousandth of a second, is shot upward.

In something like one thousandth of a second it plunges a hundred-odd miles into space and returns to a special receiver on the ground, accurately recording the height from which it was in effect reflected — the height of an ionized layer. By this method the main reflecting layers have been probed and checked time and time again. In the daytime there are normally three layers: one (E layer) at about 60 miles (100 km.); a second (F. layer) at 120 miles (200 km.) or so; and a third (F2 layer) between 150 and 220 miles (250 and 350 km.).

At night the two upper layers coalesce into one (F) layer at about 150 miles altitude, but the 60-mile layer remains in place. Rather rarely observed is still another lower layer at about 30 miles, of no great practical importance. In general, the lower (weakly ionized) layers reflect back longer waves, and the higher (strongly ionized) layers affect shorter waves.

These changes explain why a powerful broadcasting station is heard only a hundred miles or so by day, yet reaches out a thousand miles or more by night. When solar disturbances are active, some of the layer heights vary in erratic and spectacular fashion, flopping up and down by several miles in a few seconds and playing havoc with long-distance reception.

The General Picture

Before leaving the upper atmosphere it might be well to gather up some of the essential facts — the aerological elements, as a scientist would say — into one unified picture. First of all there is the matter of pressure in the air ocean above us. Pressure decreases very uniformly (but more slowly as high altitudes are reached) with increasing height. This rule never varies; pressure cannot possibly increase with height. A long and complicated formula has been deduced, called the hypsometric formula, which shows the exact variation of pressure with altitude.

The pressure at a given altitude is not always the same; it is influenced somewhat by the air temperature, and also very slightly by the amount of moisture in the air. The complete hypsometric formula must therefore include temperature and moisture terms.

In general, however, pressure decreases in the lower levels, as we have seen, by about one inch of mercury per thousand feet of altitude (or, in the metric system, by about one hundred millibars per kilometre of altitude).

At upper levels, of course, the pressure decrease with altitude is much less. At sea level the average pressure is about one thousand millibars, or about thirty inches of mercury. At about three and one half miles altitude, pressure has fallen to five hundred millibars or fifteen inches, and half of the mass of the air ocean is therefore below one. This accounts for the extraordinarily clear, transparent air characteristic of high mountain summits and high altitudes in general; it also accounts for the difficulty in breathing that most humans begin to feel at ten or fifteen thousand feet altitude. At ten miles altitude the pressure fraction has fallen to a mere tenth; at twenty miles, it is something like a hundredth; at thirty miles, something like a thousandth.

Temperature, unlike pressure, varies with height in very erratic fashion. Even in the troposphere, temperature may decrease with increasing altitude, it may remain constant, or it may even increase as one goes higher.

On the average, though temperature does decrease with height in the troposphere, by the amount of about 3° F. per 1000 feet or about 6° C. per kilometre. But above the troposphere, throughout the lower stratosphere, temperature is fairly constant regardless of altitude. This isothermal layer extends upward to about fifteen or twenty miles in winter; and perhaps extends higher, to about twenty-five or thirty miles, in summer.

Above this isothermal layer, temperature increases until, at heights around thirty miles (maximum absorption of ultra-violet by ozone), the almost vacuous air is probably as warm as the tropics at sea level (say around +85° F. or +30° C.), or perhaps up to near-boiling heat (say around +160° F. or +70° C.). Above this warm layer there is perhaps a cooler one. Then, at still greater heights above sixty miles (maximum absorption of ultra-violet by atomic oxygen), temperature may increase to several hundred degrees — the heat of a roasting oven — at least in the daytime. But at night these far heights are probably much colder — say below zero F. And above this upper warm zone, at altitudes of two or three hundred miles, for all we know, the temperature may decrease again to something near absolute zero.

The composition of the upper air has long been a bone of contention. Most meteorological texts printed prior to 1930 or so show the atmosphere above sixty miles (one hundred kilometers) as consisting largely of hydrogen and helium. More recent theoretical views of the matter, such as those of Chapman and Milne, Gutenberg and Haurwitz, have disproved this idea completely.

In the light of present knowledge it appears that the composition of the atmosphere up to at least thirty miles, and probably up to sixty miles — so far as the proportions of various gases are concerned — must be very much like that of air in the lower altitudes. At lower levels, in any case, the air consists on the average of about 78 per cent of nitrogen (by volume), 21 per cent of oxygen, nearly 1 per cent of argon, 0.03 per cent of carbon dioxide, 0.01 per cent of hydrogen, and even more minute fractions of neon and helium.

At lower levels there is also, of course, a variable amount of water vapour, ranging up to about four per cent of the air in volume — a factor of great practical importance in tropospheric weather which will be dealt with at some length. In general this water vapour decreases with height, being almost absent from the stratosphere; but even up into the ionosphere, there are probably minute traces of it.

In the upper stratosphere the proportion of helium may increase somewhat. In the ionosphere the oxygen percentage may decrease, most of what oxygen there is takes the monatomic form, and the percentage of nitrogen probably increases. Helium and hydrogen are probably not present in these black depths above the blue, warmed to roasting heat by the super-glaring sun — for if present, they would soon diffuse away into cosmic space.

Chapter 2

Pageant of the Troposphere

A towering cumulus cloud-turret depicts in clear and beautiful fashion the limpid depths of the troposphere. Slope after slope of tightly woven vapour, piled high in the ocean of clear air, catches every nuance of sunlight and shadow. Among the cloudmountains, could you imagine yourself looking straight down instead of straight up, are chasms more awe-inspiring than any abyss of earth.

For the cloud-cliffs themselves underhang one beyond the other into precipices too high and steep for belief, and beyond them are blue depths and black depths into which the swiftest plunge would find no bottom. In the hot and humid afternoon a single thunderhead may rear its enormous and opaque bulk up through the troposphere to a height of five or six miles, marking off in one perpendicular sweep the whole altitude of Everest. Above this even are the wispy mares' tails, 'locks of the approaching storm,' delicate cirrus clouds of ice crystals infinitesimal and innumerable, floating six or seven miles above the earth.

But at much lower altitudes in the troposphere — even at a mile or two above sea level — any ascending human begins to notice some shortness of breath.

For the air is rarer as one goes higher — or, to put it more scientifically, pressure in the air ocean decreases with altitude; starting, of course, from something like its average sea-level value of about fifteen pounds per square inch or thirty inches of mercury or one thousand millibars. In the lower tropospheric levels this pressure-decreasewith-altitude is quite uniform — at the rate of about one inch of mercury for each thousand feet of ascent.

As to the invisible but nonetheless real fluids which fill the troposphere, suffice it to say for the present that ordinary outside air consists of about one fifth (by volume) of life-giving oxygen, nearly four fifths of inert nitrogen, minute fractions of argon and carbon dioxide, and mere traces of certain rarer gases. Last but by no means least among the aerial fluids is a small but variable portion of invisible water vapour.

Often only a mere trace, and never more than two and one half per cent of the air by weight (or four per cent by volume), this hidden water vapour nevertheless plays a decisive rôle in weather variability and violence. For it can condense with dramatic suddenness, when aerological conditions are favorable, into fog and cloud, rain or snow or storm.

Aviators have found that by and large they can climb to about twelve thousand feet altitude without suffering any noticeable ill effects from lack of oxygen and reduced pressure, although even in summer it is likely to be uncomfortably cold at this height in an open plane. From twelve thousand to about twenty thousand feet a healthy pilot can still fly his plane after a fashion, but he will be much better off if he is inhaling a certain amount of pure oxygen (along with the thin air of those heights) from a small tank carried in the plane.

Above twenty thousand feet altitude most humans, even with artificial oxygen, find any sort of work (such as operating airplane controls or machine guns, taking photographs, reading instruments, and the like) a very arduous and wearying business.

Hence the upper tropospheric levels between twenty thousand and about thirty thousand feet are somewhat optimistically called, in aviation parlance, the 'substratosphere.' What is it like in the substratosphere?

By way of partial answer, come along with me on a substratosphere photographic flight at sunset above the city of Washington — a flight for which, being overbusied with the rather hectic activities of a rapidly expanding Air Corps, Pilot Herb and I had no time properly to prepare. But our — or my — lack of preparation at least served to make the incident more exciting.

On this clear warm evening in June the ground-level temperature in the District of Columbia was 83° Fahrenheit. The atmospheric pressure at old Bolling Field, lying practically at sea level along the Anacostia and Potomac Rivers, was of course about fifteen pounds per square inch — the usual crushing load that we dwellers at the bottom of the air ocean carry through most of our lives without a thought, because it is exactly balanced by our internal pressure.

Sweating profusely in our heavy fur-lined clothes, we clambered into *F-11*, the Fairchild photographic plane, gave her 500h.p. motor a final warming-up, taxied out to the runway, and took off at 7.05 P.M. — only thirty-three minutes before the time of sea-level sunset. Soon we were climbing steeply at about twelve hundred feet per minute (equivalent of an elevator moving straight upward at nearly fifteen miles per hour).

Herb had planned to tell me about regulating the oxygen supply during the first few thousand feet of the climb. But the full-throated roar of the motor drowned most of his words. At twelve thousand feet we gave up the attempt to talk, and I turned the oxygen valve to 'on' and put the 'pipe' in my mouth. There was too much of the heady gas for this altitude; farther aloft there would be too little. But as I did not know how to work the regulating valve, there was no help for it.

F-11 climbed more slowly and more laboriously now, at about four hundred feet per minute, as the altimeter needle crept up to 16,000, 17,000, and 18,000. The blinding orb of the sun slowly approached the horizon. It had set for earth-dwellers more than a quarter of an hour ago, but up here in the heights it would disappear some twenty minutes late. We were now just above the top of the haze layer. Sun and sky shone forth in ethereal splendor.

As the sun touched the top of the sea of horizon haze the altimeter needle touched 19,000. Our wide photographic window — a large section of the port-side wall — was not open yet. But I wanted to catch the exact sunset, and shot several pictures through the glass window — slow snapshots with the lens wide open.

The camera caught with fair verisimilitude the scene that stretched to westward. The distant ground was already shaded in virtual darkness. Just visible for fifty miles or so were the looping curves of the upper Potomac. Below the setting sun, in serrated black silhouette, were the main ridges of the Alleghenies, about one hundred miles away. Beyond these, from two hundred to three hundred miles away, was a great layer of lower clouds over Ohio, which reflected the level sunlight like the waves of a liquid sea. The haze line showed above this lower cloud layer; and above the haze, above our height even, were some thin cirrus clouds near the base of the stratosphere. The outside temperature here at nineteen thousand feet was only 28° above, and the atmospheric pressure had dropped to seven pounds per square inch — more than half the earth's atmosphere was below us.

The whole horizon, from west to north, blazed bright yellow, and even the last greenish flash of the expiring sun was bright enough to hurt the eyes. Directly beneath our feet Washington was now a hand-breadth pattern of golden threads on a background of dark velvet. Beyond the map-like confines of the city other lights were beginning to appear.

In the east the domed earth shadow was imperceptibly lifting like the bulge of some monstrous eclipse, and within it the full moon, just rising, loomed like a dim orange balloon through the lowerlevel haze. Under the moon gleamed all of Chesapeake Bay -a hundred-and-fifty-mile stretch of water from north to south that now looked like a small, long, gray puddle. The upper eastern sky was a beautiful shade of deep, dark blue.

It was time to take out the door-like left window and begin serious photography. With some effort I lifted the frame clear and peered down, straight down, through four miles of thin air. *F-11*, under Herb's experienced guidance, continued her steady climb unconcerned, but her thermometer showed an upward-decreasing, and hence unstable, temperature gradient, and the good ship was beginning to bounce around a bit in roughening air rather unusual at these great heights. The blast of freezing rarefied air pouring in the open window cut my uncovered hands to the bone. It was getting hard to read the exposure meters in the deepening dusk, except with the dubious aid of a pocket flashlight.

As the plane circled very slowly or straightened out broadside to the sunset, I fired away alternately and carefully, though very short of breath, with the still camera and the movie camera, getting my horizon level and attending to all the other things that a photographer takes care about. The earth seemed a limitless distance below, and, strangely enough, this great stretching-out of space beneath me had a disquieting effect. Worried about accidentally dropping the heavy movie camera overboard, like a non-explosive but nonetheless lethal bomb, not impossibly through the roof of the White House. Inadvertently falling overboard myself through the wide window. My hands were numb from cold, my brain was dull and wearied from oxygen deficiency, and it seemed doubtful whether, if I did fall out, I should have sense enough or strength enough even to pull the ripcord of my parachute.

It was now, at 25,000 feet, too much effort to hold the cameras properly, or even to crouch down far enough so that my line of sight on the fading afterglow could clear the upper sill of the window. I dropped a delicate exposure meter on the floor of the cabin, doing it no particular good. Rather than lift that heavy movie camera up to eye level any more, I fired some bursts of film 'from the hip' — very haphazardly in the general direction of the sunset.

Washington was now scarcely more than a fist-width web of tiny silver threads spun on darkness. I sucked hard on the oxygen pipe, but still the grogginess increased as (so Herb tells me) the altimeter needle wavered around 25,500 feet. *F-11* continued her climb at a slow, staggering gait, gaining scarcely more than a hundred feet of altitude per minute. The outside temperature was now only 14° above, and the atmospheric pressure was down to six pounds per square inch.

I had ceased now to care whether I dropped either camera on

unprotected heads below, had ceased to care whether I fell out myself, had lost interest in sunset beauties in general and this flight in particular. It was difficult even to keep my teeth closed on the oxygen tube, and to drop it here would probably mean quick oblivion. As every unnecessary motion was getting to be too much effort, I abandoned all thought of further photography, drew on my heavy gloves, and began trying to get the side window back in place before the cabin got any colder. It was no go. I couldn't even lift the five-pound window up to the opening.

My state now was much like that of an angry, impotent drunk. I was furiously mad at my own inability to do even the most essential things, and finally enraged at everything in general -at the plane, at Herb, at the idea of being up here in such a ridiculous situation — and swearing steadily at anything and everything. I was also pounding the frost-bitten fingers of my left hand on one of the fuselage longerons in order to restore some circulation — pounding at a couple of beats a second, as I thought, but actually (as Herb tells me) at more like one beat in three seconds, so markedly was the time-machine of the mind slowed down.

And then, with the altimeter needle wavering at 25,900 feet, it happened. The oxygen pipe dropped from my relaxing mouth, and fell somewhere on the dark floor of the cabin!

At this point Herb expected to see me collapse immediately. Hence he quickly throttled the motor and nosed down towards lower, denser, and warmer air. But by some strange and almost subconscious volition, I continued a ridiculously slow and stupid search for that oxygen tube on the dark floor of the cabin. In one or two minutes, just as things were going black in earnest, I somehow found the tube and got it back in my mouth.

Slipping steeply downward at half a mile per minute (measured vertically), we were soon at 18,000 feet and just above the top of the haze. Here the air remained marvelously bright and clear, but it was much warmer, easier-breathing, quieter and smoother than at the five-mile perch we had just left. In contrast, it actually seemed a rather low altitude! So we leveled off at 18,000 and prepared to watch the dying afterglow in something like serenity and comfort.

All along the northwestern horizon there was now a band of deep red that was almost purple, shading off upward to paler orange, to soft twilight white, and finally to deepening blue near the zenith. We watched this reposeful pageant, fascinated, for perhaps a quarter of an hour as *F-11* maintained her altitude in a very slow and gentle bank and turn to the left which took her around a league-wide circle.

Then, rather suddenly, I saw for the first time that rare heavenly apparition known as the zodiacal light. But such a zodiacal light! Not at all the pale glimmer of 'false dawn' you can occasionally spy by watching keenly from lower altitudes when the sun is an hour and a half below the horizon.

This was something entirely distinct from the general twilight afterglow — a localized and rather brilliant band of diaphanous white pointing up along the ecliptic for about thirty degrees, and perhaps fifteen degrees wide where it merged into the generally lighted horizon. And it seemed, if anything, rather petal-shaped — almost the way the sun's corona looks during a total eclipse. Could it, just possibly, have been part of the corona itself?

Now the full moon, climbing the eastern sky, was a ball of burnished silvery metal unbearably bright — too bright, by far, for any continued direct gaze at this 18,000-foot altitude.

As the zodiacal apparition faded, Herb throttled the motor again and pushed *F-11's* nose steeply downward. The rate-ofclimb meter swung to minus 3000 feet per minute, the air-speed needle swung far to the right, as we dove toward the expanding light-pattern of Washington, which continually grew larger and clearer and brighter below. There was the dark gleam of two rivers meeting, and the airport boundary lights standing out separate and distinct. Circling in over the dully gleaming water, our wings lit by the increasing glow of the city, we glided silently and smoothly down to a landing at 8.50 P.M. in the warm darkness. This sea-level air felt thick on our faces, and almost as hot as the blast of a furnace.

So much for the feel of the substratosphere or, more correctly, the upper levels of the troposphere. More typical of the whole great troposphere and its ever-changing pageant of cloud and clear, sunshine and storm, are the middle and lower levels, at altitudes from perhaps three miles down to the ground or the sea — levels in which weather really forms and changes and dissolves into nothingness again, weather at its best and at its worst.

In these true tropospheric levels turbulent winds rise and fall. Clouds form and dissolve; rain and hail and snow appear, freeze or melt, and perhaps evaporate again before they can lash the ground. Through small ranges of this cloud region temperatures may be as mixed as the winds and humidity may vary widely; but in general the heights are colder, temperatures decreasing on the average about 3° F. with each rise of a thousand feet (6° C. per kilometre). And as a natural corollary, the far upper levels are mostly drier than lower air layers.

But if we follow the decreasing temperature gradient up high enough — to altitudes around seven miles or so — we find that, just above the upper cirrus levels, a very curious change occurs. No longer does the temperature decrease with height; it remains constant through some miles at something like 80° F. below zero (about - 60° C.). The up-and-down winds have steadied into wide drifts, perhaps the trade winds of the future for 'argosies of magic sails, pilots of the purple twilight, dropping down with costly bales.'

Here in the stratosphere the familiar air that we breathe is rarefied almost beyond recognition, and less than one tenth of the whole mass of the earth's atmosphere remains above. Never does rain or fog obscure the moon or the stars, and the sun blazes day long in the stainless dome of the sky.

The stratosphere! What a place to explore, not hidden in foreign lands at the ends of the earth, but lying full in present view of everyone who cares to look up on a clear day; as silent and mysterious as the depths of the sea and far more bright and pure, contaminated by no continual rain of dead water creatures, but constantly bathed in the soft fire of heaven!

CONQUEST OF THE STRATOSPHERE

Not so very long ago the stratosphere was efficiently and thoroughly explored by scientists for the first time. As a clear morning of May, 1931, dawned on Augsburg, in Germany, an enormous yellow balloon shot up to meet the rising sun. From the flabby, pear-shaped gas bag hung a sealed aluminum ball, and inside it were two Belgians — Professor Auguste Piccard and his assistant, Charles Kipfer.

The take-off was so unexpectedly violent that some of the instruments were damaged, and one or two small leaks were opened in the gondola itself -leaks which had to be carefully stopped with oakum and vaseline. Our balloon rose very rapidly and in a short time we had the feeling of having left the earth behind us. It looked like a strange planet. Our altimeter announced ghastly heights of 42,000, 45,000 and 50,000 feet.

In twenty-eight minutes, climbing uncontrolled at better than interceptor-plane speed, Piccard's balloon reached its first 'equilibrium pause' at an altitude of more than nine miles. The moisture breathed out by the aeronauts had meanwhile condensed into frost on the gondola roof, and now, loosened by the intense solar radiation, showered down on them like a miniature, internal snowstorm.

After coping with these difficulties and mending new gondola leaks, Piccard dropped ballast and rose half a mile higher to a world's altitude record of about fifty-two thousand feet. At their greatest altitude the sky above the aeronauts appeared deep, dark blue. Inside the ball a thermometer read 106° F. above, while another, outside the window, showed 76° below zero. The outside pressure had dropped to one tenth of its sea-level value.

We seemed to be in an endless infinity of blue. The Earth had completely disappeared. It was like floating at sea with blue water all around and a blue shore in remote distances. The Moon appeared much brighter than from the Earth.

As the intense stratospheric sun rose higher in the dark, blue sky, a new difficulty appeared. The top of the gondola had been painted black in order to absorb enough heat for comfort; now it was absorbing far too much heat for safety! The hoar-frost melted and trickled down the sides of the gondola, and streams of sweat trickled down the faces of the aeronauts. The sunny side of the gondola became unbearably hot to the touch.

And then, to cap it all, the gas-release valve stuck. This merely meant that, for all the aeronauts then knew, their chances of getting down out of the stratosphere before dying of foul air, thirst, or starvation were at best very slim. Buoyed up by the increasing 'super-heat' or temperature-lift of the balloon under the glaring high-altitude sun, Piccard and Kipfer were prisoners in the stratosphere!

And so they remained, hung on high, a mere yellow speck as seen from the ground, for many long hours — discarding clothes, patching recurrent gondola leaks, hoping the oxygen supply would see them through — and, cool, serious explorerscientists that they were, reading the all-important instruments.

And just to enliven further those waiting hours, they managed to spill some mercury on the floor of the gondola, where only a perilously thin coat of varnish kept the active metallic liquid from eating its way out through the bottom! But after sunset, slowly cooling and contracting, the great balloon began gradually to sink. Finally, in the crisp, starlit darkness of a clear mountain evening, it made a rough and uncontrolled landing in the Innsbruck region of the Alps.

True, Piccard was not the first human being ever to penetrate the stratosphere. That honour goes to others who fought a far more difficult physical battle with cold and altitude-giddiness -aeronauts such as, for example, Captain Hawthorne Gray, of the United States Army Air Corps, who twice sounded in an open balloon to heights above forty-two thousand feet, only to meet an untimely death from oxygen deficiency on the second record-breaking ascent.

But Piccard solved at one stroke the problems of human existence in the stratosphere. He would have none of those oxygen helmets and toasting-suits, those cold and lonely struggles for breath. He fashioned an airtight ball light enough to go aloft, and opened a new era in aeronautics.

Nevertheless Piccard, more of a pioneering scientist than a practical balloonist, had his troubles and very real dangers. Piccard has ascended higher since, and so have others; women even have entered the stratosphere. Of these other stratosphere attempts, we are here concerned with only three American expeditions, in one of which it was my good fortune to be included as a forecasting aerologist.

In July, 1934, the United States Army Air Corps-National Geographic Society Stratosphere Expedition was ready to attempt the far heights. The balloon, far larger than any ever previously built by man, would leave the ground in loosely hanging folds under a round bubble of hydrogen gas in the top, but in the thin air above sixty thousand feet it would swell into a great white sphere nearly two hundred feet in diameter.

The eight-foot gondola, in which Major Kepner, Captain Stevens, and Captain Anderson were to go aloft, held more scientific equipment than had ever been assembled in such a narrow space before, and resembled nothing so much as the interior of a submarine. There were two large and several small cameras, apparatus for measuring cosmic rays (those minute messengers from the depths of space that may hold the secret of creation), sampling flasks to capture some of the high atmosphere, barometers, thermometers, and altimeters of various types, spectrographs for recording the spectra of the sky at various angles, heavy electroscopes shielded with lead, a complete air-producing and air-conditioning outfit, a complete radio receiver and transmitter for two-way communication, and no end of batteries to supply current for all the complicated apparatus.

Weather favorable for the flight appeared on the twentyseventh of July. The great balloon was inflated during the night in its natural bowl in the Black Hills of South Dakota, sheltered by 400-foot cliffs. Shortly after five o'clock in the morning the last holding ropes were released, and the ponderous fabric floated clear of the Bowl and of Mother Earth. Leveling off at 5,000 feet to complete the disposition of outside apparatus, and again at 40,000 feet for scientific observations, the three aeronauts continued their sounding into the stratosphere.

We stopped valving gas [writes Captain Stevens] * and the *Explorer* again started to ascend. It was nearly noon, mountain time, and for the next hour we moved steadily upward. At 1 o'clock we were approaching the 60,000-foot level, and Major Kepner prepared to bring the balloon again into equilibrium. Valving was begun, and the balloon started to slow down. . . .

At the time, I was giving some instrument readings over the radio. Suddenly a clattering noise was heard on top of the gondola.

We looked upward through the three-inch upper port and saw that the noise had been caused by part of the appendix cord -a small rope — falling on the roof of the gondola.

What had caused the cord to drop? Looking still higher we were startled to see a large rip in the balloon's lower surface.

Through the overhead glass porthole we watched the rent in the fabric gradually becoming larger and larger. The minutes passed slowly by; the magnets of the cosmic ray instruments clattered on; the buzzers hammered on the barometer box; the instrument cameras clicked in unison at regular intervals.

At times we all talked briefly over the radio. But little other talking was done, for our ears were strained to get warning sounds from above us. Soft swishing noises came through the roof of the gondola from time to time. Each of these sounds meant a new rent, or an increase in length of a rip already there.

Below us was the brown, sun-baked earth, so far away that no roads, railroads, or houses could be made out. Our direction of drift was changing, but that was now a matter of little concern. The question now was not where we should get down, but how!

As we looked through the ports that were 45 degrees from the vertical, the sky was the rich dark-blue colour associated with high mountain views; but from the vertical port the sky was like black velvet on which ink has been spilled and dried — it was black with just the merest touch of dark blue.

But we could not fail to note the astounding brilliance of the sunlight and the intensity of the reflected light from the balloon rigging. Some of the ropes, especially those nearest the gondola, were so bright that they had a fluorescent appearance and looked larger than they really were.

From our instruments we knew that outside it was almost 80 degrees below zero, Fahrenheit. We were in relatively warm air inside, about ten degrees above freezing, but it was getting steadily colder. Already ice had formed in our gondola in a narrow band about two feet below the top.

Three quarters of an hour passed, and we were down to 40,000 feet. Our speed was increasing, and half an hour later we were down to 20,000 feet. Major Kepner and Captain Anderson each forced open a hatch, and for the first time we felt we were free.

We all climbed out on top and took a good look at the balloon. It was pretty badly tom. Many more tears and rips had appeared in it. The question was, how long would it hold together? As the enormous bag came downward through the air, large waves appeared in the lower fabric, sweeping across it and back again. With almost every wave the rents grew bigger and bigger. Suddenly the entire bottom of the bag dropped out. We could look up into the whole bare inside of the balloon. Only the part above the lower catenary band now remained. The bottomless bag was acting largely as a parachute. It was a pretty sight, quite round and tight and symmetrical. But it was a bit too tight for safety!

In the meantime Captain Anderson, atop the gondola, had been having difficulty with his parachute. The release handle had caught on something and the parachute pack had come open. It was a situation that might have been disconcerting to a less cool head. There was only one thing to do — that was to gather the folds of silk under one arm preparatory to leaping. While getting the fabric together, Anderson stepped down until both his feet were in the hatch from which I planned to leap. Whether Anderson heard me or not does not matter. Things started to happen fast. The feet disappeared, and I knew he had leaped. As he jumped, the balloon exploded.

The metal gondola, plummeting downward at something like a hundred and fifty miles an hour with its heavy load of scientific equipment, crashed into the black dirt of a Nebraska wheatfield and was totally destroyed.

The men behind this stratosphere expedition were not to be deterred by one failure, however destructive and dangerous. May, 1935, saw the camp again established in the Stratobowl. Most of us in that phase of the expedition had arrived by airplane from the East, a new and even larger balloon was being completed in Ohio, and a new nine-foot magnesium-alloy gondola was ready to house even more scientific instruments than before.

As hydrogen gas had been chiefly responsible for the violent explosion of the balloon in the preceding year, hundreds of thousands of cubic feet of expensive helium had been provided. The balloon fabric alone weighed some seven thousand pounds; the loaded gondola, thirty-five hundred pounds; lead-dust ballast, six thousand pounds; and the total lift (including a 'free lift' of six hundred pounds for the take-off) amounted to nine tons!

Early in June all preparations were complete, and there began the long, tedious wait for the weather, of which, in great variety, the Black Hills have a plentiful share. For ten days it rained and snowed almost continuously. During one whole night a great thunderstorm growled and flickered just west of us, the most violent I have ever witnessed, within the tropics or without them, and the hail it spilled down was reported to be as big as baseballs. On our camp it ranged from large-marble to small-golf-ball size.

The peculiarly favorable weather conditions required for the success of a stratosphere flight naturally operated to make 'inflation days' few and far between on the calendar. First of all, we must have very low winds and no rain on inflation day for the laying out of the great balloon in its bed. Then, during the whole of 'inflation night,' when the towering fabric of the great balloon was rising slowly to its full height of three hundred and twenty feet, like the gradual growth of a gigantic mushroom, we must have very low winds on the rim and flat calm in the Bowl.

On the morning of 'flight day' we must have very low winds for the take-off; and throughout flight day, over the whole broad expanse of South Dakota and Nebraska, we must have skies either completely clear or blemished only by a minor fraction of cloudiness. Finally, we must have low winds for landing the great balloon a few hundred miles to the eastward. Is it any wonder that we had to wait?

But at least the Black Hills offered no end of interesting hikes and climbs, and the people of South Dakota were more than hospitable. We amused ourselves, also, by sending up large pilot balloons which indicated the upper winds as high as eighty-two thousand feet.

After a few excitements and arguments over favorable-appearing weather situations, inflation day finally came on the eleventh of July. The great balloon was laid out in the afternoon sun, and helium gas rushed hissing into the long cloth pipes as some of the hundreds of cylinder valves were opened.

Throughout the calm night, black above the floodlight glare in the bottom of the Bowl, the slow and careful preparations went on as the Stratocamp hummed with activity. Inch by inch the great white mushroom — seven thousand pounds of woven fabric held up by a great bubble of buoyant gas — towered silently into the still air of the Bowl.

Nearly three o'clock in the chill Dakota morning, and the gondola was brought out of its shed and wheeled under the balloon. Perhaps a score of us were standing directly under the towering folds of fabric — working, watching, and talking of final arrangements. Suddenly, without the slightest warning, there was a rustling, whistling sound overhead that grew by the moment into a noise like muffled thunder. The balloon had burst, and it was falling on our heads! Everyone ran radially outward from the falling mass, colliding, falling down, scrambling up again at a full sprint. Then the fabric hit the ground, not ungently, and those who were clear turned instantly back to help those few who had been caught under acres of cloth.

In three seconds of ripping fabric and falling ropes the spring expedition of 1935, which had cost thousands of dollars and weeks of trying work, came to an end. It was a bitter moment for Albert Stevens, Orvil Anderson, and their associates. But these men were not to be denied.

October of the same year found the Stratocamp again active, with tents boarded up against the early approach of winter. Some who had been there in the summer were held elsewhere by other duties, but Stevens and Anderson were there and ready to dare the stratosphere for the third time. The great balloon had been repaired, and cylinder banks of new helium were spotted in the Bowl.

Despite the bitter cold, all preparations for the third attempt went swiftly and easily. Inflation day came on the tenth of November. The laying out of the balloon proceeded smoothly, as before. The cylinder valves were 'cracked,' and gas began hissing into the great balloon.

The mushroom began its slow growth; and then, to the consternation of all, a twenty-foot rip was discovered in the belly of the balloon! Here was a situation that called for the display of the qualities that have made the Anglo-Saxon race and tradition what they are — the hard and imperious talent to 'see it through.' The *Explorer I* had exploded over the heads of Stevens and Anderson, a mile above Nebraska the year before. This same *Explorer 11* had burst over their heads in the Stratobowl only four months before. And now, even before the inflation was complete, the *Explorer II* showed a twenty-foot rip. The rip was patched and the inflation went on.

Early the next morning, somewhat behind schedule, the balloon rose smoothly under the powerful upward pull of its helium. Despite a slight breeze that had sprung up, it was easily clearing the 400-foot rim of the Bowl, crowded with swarms of interested onlookers, when something unforeseen happened. A downdraft caught the great fabric, and it started downward toward the rim. Brief and earnest prayers, calling the deities shortly by name, went out from the gondola radio transmitter over a nation-wide broadcasting network.

Anderson threw the ballast-release switches, blasting out bag after bag of lead dust, which poured in streams on the heads of the scurrying multitude below. Within fifty feet of the treetops the great balloon stopped her fall, paused momentarily, and then started upward — off for the stratosphere and an enduring world's altitude record of 72,395 feet. It may sound ridiculous but to me the only moment of the entire flight that was breath-catching was the time when the big central appendix first opened at 65,000 feet and through it I saw the dome of the balloon so far, so very far, above us. It was incredible that we were riding under such a mighty ball of gas!

Observations of the earth and sky were made at the top of the flight, 13.71 miles above sea level.

A photograph taken at maximum altitude — the highest photograph ever taken — shows the section squares of western Nebraska, the Dakota Bad Lands, and the Black Hills. Moreover, this photograph clearly shows, for the first time, the round bulge of the horizon on this our great ball of an earth.

The earth could be seen plainly underneath through the lower porthole and hundreds of miles in every direction through the side portholes. It was a vast expanse of brown, apparently flat, stretching on and on. Wagon roads and automobile highways were invisible, houses were invisible, and railroads could be recognized only by an occasional cut or fill. The larger farms were discernible as tiny rectangular areas. Occasional streaks of green vegetation showed the presence of streams.

Here and there water could be seen in the form of rivers or lakes, especially if the sun was reflected from the water's surface. No sign of actual life on the earth could be detected. To us it was a foreign and lifeless world. The sun was the one object that com- manded our attention; we were temporarily almost divorced from Mother Earth. To look directly at the sun through one of the portholes was blinding. The sun's rays were unbelievably intense.

Through the upper porthole we looked often at the vast bulk of the balloon, upon which the sun beat fiercely. Through the central appendix we could look up to the very top of the balloon. The two valves, almost a hundred yards above us, appeared like little buttons. The tapes of the massive dome formed a perfect pattern — just like the meridians of longitude and parallels of latitude on a globe of the earth. Overhead, the great balloon blocked our view of the sky above us. How I wished at this time that we could have a central tube in the balloon through which we could look at the zenith! I am sure that the sky would have been so dark directly overhead that we could have seen the stars at noonday.

As it was, we could see the sky, beyond the sides of the balloon, at an angle of about 55 degrees above the horizon. The horizon itself was a band of white haze. Above it the sky was light blue, and perhaps 20 or 30 degrees from the horizon it was of the blue colour that we are accustomed to. But at the highest angle that we could see it, the sky became very dark. I would not say that it was completely black; it was rather a black with the merest suspicion of very dark blue. Thus the violet sky, which theory formerly postulated for the daylit stratosphere, seems to fade into the limbo of onlyimagined things.

Listening at a radio near Washington while Stevens was at his summit over South Dakota, I distinctly heard him say: 'Still a blue sky, but a very, very dark blue — directly overhead. Previously we thought it was a black sky, but it's undoubtedly blue.' But perhaps if he could have looked straight up towards the emerging stars, there would have been some tinge of violet. Perhaps in the lower ionosphere, far higher than any balloon can go, the sky glimmers violet for a space before it fades to velvet blackness.

PILOTS OF THE PURPLE TWILIGHT

Sounding balloons carrying self-recording instruments have gone more than twenty miles aloft. They have shown us that temperature in the upper stratosphere tends to increase gradually with increasing altitude. At the stratosphere base, or 'tropopause,' it is perhaps on the average around 80° F. below zero (about — 60° C.). At still greater heights, the thin air warms up, relatively speaking, to only 20° 0r 30° below. Sounding balloons have also told us that the stratosphere base has a gradual slope from equator to pole. Over the tropics it is about ten miles high, over temperate latitudes around six or seven, and over the poles only four or five. Moreover, the stratosphere base over the equator is considerably *colder* than the stratosphere base over the poles. Thus paradoxically, the coldest place on earth (around — 115° F. or 80° C.) is perhaps twelve miles above tropic lands and seas.

However valuable they may be in the preliminary pricking of the stratosphere, balloons will never thread it through with the air lanes of commerce. And the problem of climbing ten miles or so on wings is far more difficult. The wings themselves must be longer and narrower than those used for ordinary flying. For comfortable and regular stratosphere navigation there must be a strong, airtight cabin, supplied with artificial atmosphere by an air-compressor; and from this cabin the flying controls must be operated with certainty and ease.

The motors, driving variable-pitch, large-diameter propellers that will get some purchase on the thin upper air, must also be fed something like a sea-level, oxygen-nitrogen mixture by one or more air-compressors or 'superchargers.' At seven miles, where the stratosphere begins in middle latitudes, the atmospheric pressure is about one fourth of its sea-level value; at twenty miles, perhaps the upper limit of the stratosphere as we know it, the pressure has diminished to a mere one hundredth. Between ten and twelve miles, where lie, perhaps, the high-altitude air lanes of the ultimate future, pressure fractions range from one tenth to one twentieth.

While there had been talk of substratosphere and stratosphere airplanes from the time of the First World War, serious American research in this field really began in 1937 when the Lockheed Company delivered to the Army Air Corps an experimental substratosphere ship called *XC-35* — a fast, transport-type, low. wing, bi-motor monoplane of about sixty feet wingspread.

Both of *XC-35's* 500-h.p. radial motors were supercharged (or fed compressed air by centrifugal-type compressors) so that they would function efficiently at more than five miles altitude. Within the sleek, aluminum-alloy fuselage was a sealed pressurecabin large enough to enclose a maximum crew of six men, flight controls and instruments, and various experimental apparatus.

Whatever the altitude reached by the stratoplane, cabin pressure could be maintained by blower-compressors at about 10 lbs/in^2 (which is equivalent to the free pressure of the open air at 12,000 feet altitude). With the stratoplane cruising at 30,000 feet altitude, the pressure differential between inside and outside would be about 6 lbs/in^2.

But for added safety, the pressure cabin was made strong enough to stand an internal pressure of 15 lbs/in^2; and the whole structure was made flexible enough to accommodate itself to any temperature between +100° F. and -54° F. The cabin supercharging system also took care of ventilation, pumping in fresh air and removing stale air. Even stratosphere air, despite its low density, contains the same oxygen percentage as sea-level air; hence no artificial oxygen was ordinarily needed in the Army stratoplane. Nevertheless it was carried as an emergency reserve.

On a test flight in January, 1938, *XC-35* cruised over a 220-mile course at altitudes between 19,000 and 20,000 feet, carrying Captain Johnson (pilot), Captain Sims (co-pilot), and two observers. The average speed-at-altitude was around 350 miles per hour, or more than one third again better than the lower-level performance of the same plane.

Finally, in 1940, American transport aviation really began to move 'upstairs' in its unrelenting search for improved speed, comfort, and safety. During the summer of that year there appeared on one of our transcontinental airways (T.W.A.) the Boeing 'stratoliner' — the first regularly scheduled and flown supercharged-cabin transport airplane in the world.

The 'stratoliner' was a beautiful big monoplane with 107-foot wingspread and a 75-foot cigar-shaped fuselage nearly twelve feet wide, all fashioned of brightly gleaming aluminum alloy. Four radial air-cooled engines, each developing over a thousand horsepower, were cowled into the leading edge of the single low wing. The entire oversized fuselage, from the roomy forward control cockpit or 'flight bridge' back to a pressure bulk- head near the tail, was scaled and strengthened to a maximum pressure differential of 6 lbs/in^2. Each of the four engines was supercharged for efficient high-altitude operation, and cabin ventilation and supercharging was accomplished by two complete duplicate systems, either of which could on occasion handle the whole load alone.

The total weight of this revolutionary substratoliner was slightly over twenty tons. It could carry more than thirty passengers in addition to a crew of five through the upper levels, beyond reach of ordinary low-level storms, at an easy cruising speed around 210 miles per hour.

At 20,000 feet altitude — the highest routine flying level contemplated in the original design — the cabin would be supercharged to a pressure differential of 2½lbs/in^2, so that passengers lolling in yacht-like comfort might breathe a warmed and conditioned atmosphere at about two thirds of sea-level pressure. By August, 1940, the new substratoliners began making the 2400-mile transcontinental run in a scheduled time of about thirteen hours, and sometimes made coast-to-coast in twelve hours or less.

All this substratosphere flying is, of course, far below the ultimate ceiling, or maximum possible altitude, of the airplane, and far short of ultimate stratospheric speeds. For the conventional airplane as we know it today (ruling out for the present, that is, such possible ultimate developments as rocket ships and so on) — for the airplane this absolute ceiling is probably around eleven miles above sea level. Some fairly close approaches to this ceiling have already been made by military flyers in specially designed planes.

On September 28, 1936, for example, Squadron-Leader Swain, of the British Royal Air Force, took off from Farnborough, England, in *K4879*, a specially constructed monoplane. It had sixtyfoot-wide wings, built of laminated wood for extreme lightness and black-painted on their upper sides to reduce glare; a beautifully streamlined fuselage, also of wood; a powerful engine equipped with a two-stage engine-driven supercharger which enabled it actually to develop more horsepower at eight miles altitude than it did at sea level; a great thirteen-foot propeller having four wooden blades. Swain himself wore an airtight rubberized-fabric suit and metal-celluloid helmet — a sort of diver's suit for soundings far upward in air rather than downward in water — which was kept inflated to livable pressure and air-conditioned with fresh oxygen and air-freshening chemicals.

K4879 reached a world's airplane altitude record of about 50,000 feet — and more important, remained near that altitude for two hours. The upward climb itself took only an hour or so. On the long downward glide, however, something went wrong with the air circulation of the airtight suit. Luckily the highaltitude ship was then below 20,000 feet, and Swain promptly ripped away the front windows of his helmet with a knife, thereby saving himself from a strangling death. If *K4879* had been equipped with a more modern type of engine compressor -a turbo-supercharger driven by the engine exhaust — Swain might have reached still greater heights.

A considerably higher airplane altitude record was claimed by Lieutenant Colonel Pettzi, of the Italian Air Force, in December, 1938. His special stratosphere-ship was a wide-winged Caproni biplane; fitted, like the United States Army stratoplane, with a small supercharged cabin in order to defeat the lethal qualities of high-altitude air. Pettzi reached (according to the Fascist claim) an extreme altitude of about fifty-six thousand feet.

When aeronautical engineers have conquered not only the substratosphere, but also the stratosphere itself, what widening horizons they will bring us! Aerial sightseers, now content with the modest vistas of Mount Washington and Pike's Peak, may in coming years embark on two-hour altitude voyages. They will see the blue sky imperceptibly deepen toward purple or black, with here and there perhaps the pinpoint glitter of a planet or bright star. The daytime moon will stand forth luminous and marvelously clear, every detail of its sunlit surface etched in relief, as it appears on earth only in a limpid clearing sky washed clean by rain or snow.

The stratosphere tourists will have to screen their eyes from the blinding daggers of diamond shot forth by a sun too fierce for direct gaze, however momentary. Below them, plainly to be seen in clear weather, will spread the varied tapestry of their entire state, its hills and valleys flattened and dwarfed and browned, its cities distinguished chiefly by their mantling palls of unhealthy smoke, its sunlit rivers and lakes stretching away hundreds of miles to the haze-dimmed horizon.

Scientists could take up cameras and other instruments for high-altitude observations. A tropical hurricane might be charted and photographed from a safe position directly above its seething fury. A wide-angle camera lens might catch the whole broad circular shadow of the moon sweeping at three thousand miles an hour across the earth during a total solar eclipse. Acrologists might observe in the upper air changes portentous of coming weather on the earth below.

In wartime what weird and horrible bombers some of these stratosphere ships might make, flying on far missions through the cold silences of the upper levels, hidden alike from enemy and friend! Their bombing accuracy would not be very good, but their secrecy and stealth and surprise value would be considerable. As the ships arrive over their objective, charged plummets start downward, gaining speed as they fall through the thin air, and some minutes later strike home, erupting in dense clouds of black smoke from scattered half-acres of destruction. And the stratosphere bombers drone away in the quiet blue, untroubled by surface guns or ordinary pursuit planes, unseen, and perhaps unheard.

Wings in the stratosphere should find their true mission, however, as the high-speed transports of the future. Perhaps the day will come when New York and San Francisco will be only four hours apart by high-altitude air — or say three hours and fifty minutes eastward with an average fifty-mile tail wind, and four hours and a quarter westward against the prevailing aerial drifts.

Fancy among the passengers a small boy who has just come into possession of a new watch, trying to set this absorbing instrument ahead half an hour every forty minutes, in accordance with the confusing clock in *ST-7's* cabin. 'Hey, mister, we leave San Francisco at noon and don't get to New York till 7, and *ST-8* leaves New York at 3 and gets to San Francisco at 4. Say, what do you mean we travel faster than they do? Say, are those little molehills down there mountains, honest? While the tangled wisps of cirrus weave a gradually thickening veil between *ST-7* and the broad earth, her pilots set their course by radio range and listen to radiophone weather broadcasts, watching from their serene height the slow development of low-altitude storms.

As the ship nears the Atlantic seaboard, her passengers can see through the port cabin windows a spectacle of cosmic magnificence never duplicated on earth — sunset viewed from a height of ten miles, through only one tenth of the earth's atmosphere. The stratosphere passengers sit quiet, awed — to stir only after *ST-7*, sinking steadily on the long glide to her terminal, has lost somewhat of cosmic glory in the unhurried approach of innumerable lights on the dark ground below.

In a later development of stratosphere transport, may not the oceans eventually be bridged by non-stop flights within the space of a few hours? Perhaps the *AST-3* leaves London at 6 P.M., bound for New York. It is larger than the transcontinentals, its powerful radio can raise both shores from the middle of the ocean, and a navigator continuously plots its position by celestial observations and radio bearings. In case of trouble over the Atlantic, from a height of ten miles it can glide more than a hundred miles to a safe landing alongside some convenient steamer. Using rocket propulsion in the stratosphere, after its climbing propellers have been feathered into the wings, it cruises comfortably at nearly seven hundred miles per hour, making the Europe-America run in a little less than five hours.

As *AST-3* gains altitude over the lower reaches of the English Channel, its passengers begin to notice a very curious phenomenon. The sun, which seemed on the point of setting, is now arrested by some invisible but all-powerful Joshua; after an hour or two, in fact, it actually appears to be imperceptibly rising! *AST-3* is speeding westward faster than the great earth itself rolls around in its diurnal spin, and its passengers have temporarily escaped the inexorable march of time.

Out over the unending miles of blue water, broken here and there by the white wool of scattered clouds and the more extensive blankets of storm areas which are soon overpassed and left behind, the passengers amuse themselves by watching for the smoke trails of steamers on the ocean floor below. Even when directly beneath, the surface leviathans themselves look like small and slow water-crawling insects, wearily dragging diminutive wakes across ripples such as the gentlest breath might stir on the surface of a pond.

Soon on the far edge of the bright sunlit water ahead appears a narrow line, dark and incredibly long, which gradually broadens as it grows out of the western horizon. It is America, land of hope and promise, journey's end.

Chapter 3

The Sun and the Earth

The baffling complexity of weather changes on earth propounds to inquiring humans an everlasting how and why. Where does our weather in this north temperate zone, for example, truly originate? Most immediately, in some cyclonic disturbance moving towards us. Less immediately, in the push of polar air out of the frozen north, or in the flow of tropical air out of the steaming south. But what causes this ebb and flow of unlike air masses, what inspires the planetary centers of pressure-action that drive their great mechanism? We look about on this broad planet for the original controlling causes, and we do not find them; nor shall we find them short of ninety-three million miles out in space, or elsewhere than in the sun.

The Earth in Relation to the Sun

There was a time before the beginning of the earth when our sun was born, along with the other stars in our galaxy, perhaps out of the slow condensations of diffuse matter in some vast spiral nebula. Then, perhaps, as the ages passed and our sun wandered lonely in the void under the vague pull of unseen forces, it chanced to pass close to another sun also wandering in space. Gravitational attraction between the two suns raised on each enormous tides, which probably took the form of great eruptions of gas extending thousands of millions of miles out ward. These gaseous eruptions gradually broke apart, slowly cooled, and finally condensed into what are now the planets, this earth among them. Now, the earth and its moon circle perpetually in the glare of the great sun. Night and day they are bathed in its tremendous blaze of radiation. Though the earth itself receives but an infinitesimal fraction of the sun's total outpour of energy, even this fraction amounts continuously to twenty-three thousand billion horsepower. The sun is the engine that drives our winds and our weather, and makes the earth livable for men. The coal and oil which keep us warm in winter are only particles of the stored sunlight of the past, as water-power is the stored sunlight of the present. Every factory wheel turns, every radio message speeds, every boat, train, automobile, and airplane moves only by the delayed impetus of solar energy.

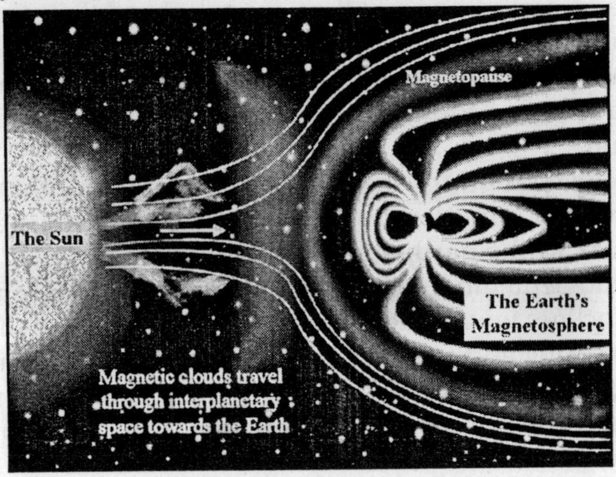

Fig. The Sun and the Earth

Perhaps we can best visualize the earth-sun relation, stripped of astronomical figures, by means of a model. The sun is a ball twenty-two feet in diameter — about the size of an average living-room. Half a mile away moves a tennis-ball 'earth'; attended in turn by a 'moon' the size of a small marble, circling its earth at a distance of about six feet.

As this tennis-ball earth circles slowly around its living-room sun, it rotates in the same direction like a ball rolling around the edge of a circular disk; and this continual rotation makes the sun, and all the sky along with it, appear to rotate from the viewpoint of microscopic creatures inhabiting the surface film of the tennis ball—gives them the regular and alternating rhythm of day and night.

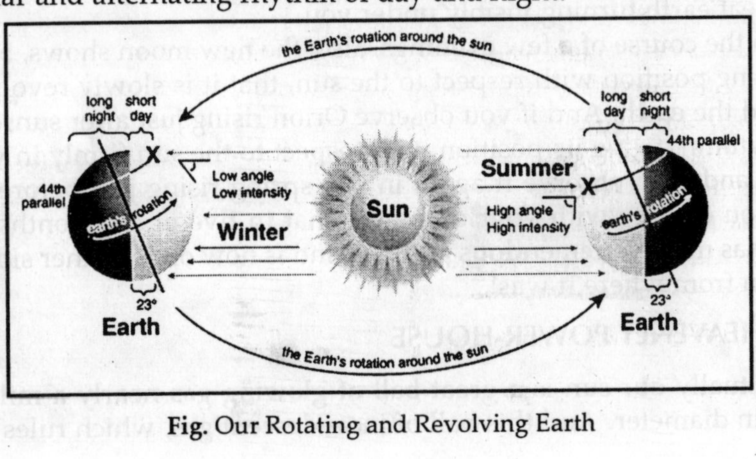

Fig. Our Rotating and Revolving Earth

If the axis of the earth's rotation were perpendicular to the plane of its orbit around the sun, the day would always last twelve hours everywhere on earth (except at the very poles) and there would be no seasons. But actually this axis is inclined twenty-three and one half degrees from the vertical, and remains pointed very nearly in one constant direction as the earth rotates on its axis and as it revolves about the sun.

This inclination brings the twelve-hour day to all the earth only twice each year, on the two days when the inclined axis is perpendicular to the earth-sun line. Between these two equinoxes (March 21 and September 21), there is a six months" season when the North Pole inclines towards the solar centre, bringing us in the northern hemisphere spring and summer with their long days and high-riding sun. Between September 21 and March 21 there is another six months' season when the axis inclines away from the solar centre, bringing us fall and winter with their short days and low sun.

Some of these relations, easy enough to understand in terms of models, can be seen by observant eyes in Nature itself. If you look at the new moon through a good pair of prism binoculars when its silver crescent is thin and the earth-light gleams faint green on the 'old' moon, the whole moon stands out in your vision as the great ball that it actually is, swinging ponderously in space.

Near the horizon opposite the sun, after sunset or before sunrise in the clear air of high altitudes, you can see the earth's domed shadow, dull gray against faint rose, show- ing plainly the curvature of our planetary globe. If you watch the setting moon or the setting sun through a fixed, moderatepowered telescope just as it nears the trees of a distant hillside, and bring yourself to imagine for a moment that the moon (or the sun) is stationary, then you can see and almost feel the great earth turning visibly under you.

In the course of a few evenings, also, the new moon shows, by its changing position with respect to the sun, that it is slowly revolving around the earth. And if you observe Orion rising just after sunset in late autumn, fixing its position with respect to the sun firmly in your mind, and then observe it again in late spring rising just before the sun, you can see with your own eyes that in five or six months the earth has made a tremendous journey, and is now on the other side of the sun from where it was!

OUR HEAVENLY POWER-HOUSE

Actually our sun is a great ball of glowing gas nearly a million miles in diameter. And this ball of incandescent gas, which rules our winds and our weather, our motions and our lives, from nearly a hundred million miles out in space, is in all reality a gigantic laboratory of physical wonders.

Although the sun is entirely gaseous, its inner layers are under such tremendous pressure that the gas is more dense than any earthly solid, and the temperature runs into millions of degrees. The temperature is so enormously high, in fact, that some of the more complicated atoms are being constantly broken down into simpler atoms.

Thus that perennial dream of ancient alchemists, the transmutation of elements, is an everyday reality within the sun. Furthermore, the later and much wilder earthly dream of releasing subatomic energy, by which an airplane might fly for months with nothing more in the way of fuel than a drop of water, is going on constantly in our great heavenly power-house — is, in fact, the only possible explanation of the sun's miraculous power to radiate, day after day and year after year, such a surprising flood of energy. Even so, the sun cannot warm us without radiating away its own mass and weight at the rate of more than four million tons a second.

Even near the surface of the sun temperatures are far hotter than any earthly furnace can produce, for the photosphere glows at about 10,000° F. or 6000° C. It is not easy to appreciate what such a temperature means in the way of power. At the photosphere, solar radiation would melt in one minute a shell of ice nearly forty feet thick.

A section of solar surface the size of a thumbnail would develop energy continuously at the rate of forty horsepower — enough to run an automobile or a small airplane; and a square foot of solar surface would drive an ocean steamer. To indicate the total horsepower radiated continuously by the entire sun, we must write the number 465 followed by twenty-one zeros!

At the photosphere, pressures and densities are comparable to those of the air we breathe. Just above the photosphere is the so-called reversing layer of somewhat cooler gas; and above this lies the chromosphere, seen, if at all, as an irregular ring of pinkish light. Beyond the chromosphere there extends out for millions of miles the faint, pearly nebulosity of the corona, seen from within the earth's atmosphere only during a total eclipse or by means of complicated optical apparatus.

If we look at the sun through a telescope (equipped, of course, with some device for cutting down the excessive light), or even a large field glass, we note that the surface darkens perceptibly towards the edge or 'limb.' By tapping the telescope one can see that the whole solar surface is granular, and that this 'ricegrain structure,' each grain of which is some three hundred miles in diameter, constantly flows and changes like the waves of a restless sea.

Over a span of some days, by watching the movement of sun spots, one can note also that the whole sun is gradually turning. The entire solar globe rotates on its axis, with higher latitudes lagging somewhat behind the equatorial belt, in an average period of about twenty-five days. But during a solar rotation the earth has moved along its orbit in the same direction, so that to us the period appears to be about twentysix or twenty-seven days.

The Sun's Radiation

Several and varied are the forms in which solar energy is radiated into space. Actual electrons, minute building-bricks of all matter, are shot off freely in all directions and uncounted numbers at speeds around a hundred thousand miles an hour. Heavier ions, each approaching the size and mass of an atom, are also showered copiously out through the solar system. But the most powerful, continuous, and important form of solar radiation consists of electro-magnetic waves of heat and light and ultra-violet.

Light (and radio waves, heat, and X-rays as well) consists of electro-magnetic vibrations, or waves, in space without any necessity for air as a transmitting medium. If we hold a glass prism in a beam of sunlight we see at once the familiar spectrum of colors which, mixed together, make white light; Nature herself reveals it occasionally in the rainbow. Each colour in this spectrum appeals to our eyes as such because the rate of vibration of its impulses is different from that of any other colour.

These vibration rates for visible light, like those of the music scale, extend through about one octave. Leaving out the lower and upper *do*'s for the moment, *re* corresponds with red, mi with orange, *fa* with yellow, *sol* with green, *la* with blue, and *si* with violet. Yet when an electric heater is turned on, though it does not begin at once to glow visibly, an unwary finger would nevertheless feel heat.

There is, then, a 'dark color' below red on the scale, known as infra-red, or simply as radiant heat; this corresponds with lower *do* on the music scale. Also, if we hold a piece of photographic film in the solar spectrum just beyond the visible violet, it will very soon be exposed, even though we cannot see that any visible light is reaching it. This other 'dark colour,' corresponding with the upper *do* of the music scale, is called ultra-violet. Each of these two kinds of dark light extends through a greater range of vibrations than all the visible colors combined.

In sound we can talk of the number of vibrations per second, such as 256 for middle C; or of the wavelength (distance between crests), which is about four feet for the same note. The number of vibrations times the wavelength always gives the speed of travel, which in the case of sound in air is about 1100 feet per second.

But in the case of light this speed of travel is so enormous (about 186,000 miles or 300,000 kilometers per second) that a curving beam would go about seven times around the world in a second, the vibration rates are so high that it is impossible to visualize them and toilsome to write them, and the wavelengths are so short that a special unit is needed to measure them conveniently. This unit, about one two-hundred-andfifty-millionth of an inch, is called the angstrom, after a nineteenth-century Swede who pioneered the study of solar radiation with the spectroscope and invented the compensation pyrheliometer for accurately measuring its total amount.

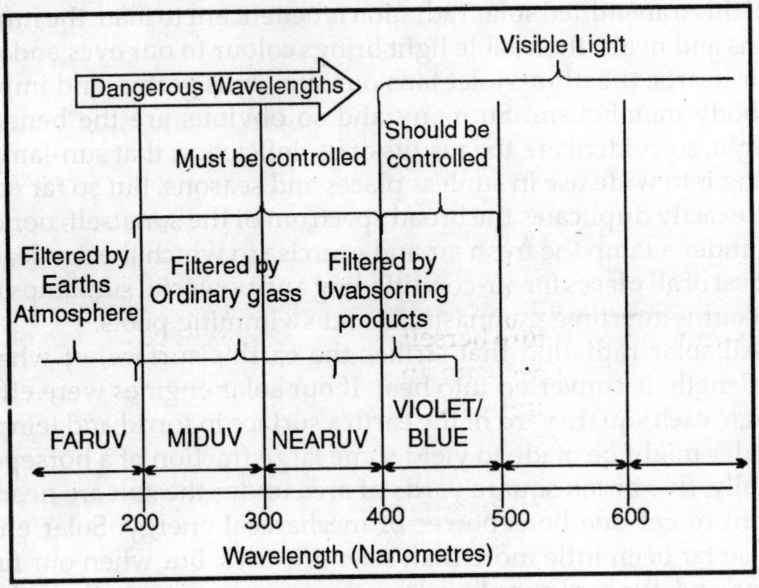

Fig. Visible and Invisible Light

In the vibration range of the visible spectrum, longer-wave red begins at about 7600 angstroms, and shorter-wave violet ends at about 3900 a. Everything below (or longer than) 7600, down to where the radio waves begin, is infra-red or heat. Everything above (or shorter than) 3900, up to the longest X-rays at about 1000, is ultra-violet.

The sun itself radiates a continuous spectrum of colors smoothly merging into one another — a broad band of wavelengths extending all the way from long heat waves up into the X-ray region. Of this total solar radiation, more than four fifths is infra-red, less than one sixth is visible light, and less than one sixtieth is ultra-violet. Yet this small proportion of ultra-violet is in some ways as important in human affairs as the longer-wave radiation.

Sunlight on Earth; Insolation

Fortunately for human beings, our earth's atmosphere does not pass the whole gamut of solar energy. For solar radiations shorter than 2900 a. would, if they reached the earth's surface, probably destroy human health and life as potently as the longer ultra-violet now upbuilds them. As it is, we are effectively screened from these destructive radiations by an atmospheric phenomenon that the solar radiation itself produces — a diffused layer of ozone in the thin air far above the earth. Given clear skies, a large part of the remaining solar radiation (from 1900 a. to 20,000 a. (or so) reaches the earth, and serves mainly to warm it. All this transmitted solar radiation is beneficent to man: the infra-red warms and heals; the visible light brings colour to our eyes and gaiety to our hearts; the ultra-violet tans our skin, kills germs, and improves our body metabolism. So many and so obvious are the benefits of sunlight, so evident are the results of its deficiency, that sun-lamps are coming into wide use in sunless places and seasons. But so far no sun-lamp exactly duplicates the broad spectrum of the sun itself, nor do we find under a lamp the fresh air and exercise to which the sun leads us. The best of all places for air-conditioning and powerful sunlamps might be in our wintertime gymnasiums and swimming pools.

All solar radiation that strikes the earth's surface, of whatever wavelength, is converted into heat. If our solar engines were efficient enough, each square yard of the earth's surface in torrid and temperate latitudes might be made to yield some large fraction of a horsepower. Actually, five or ten square yards of area under the sun are needed at present to get one horsepower of mechanical energy. Solar engines have so far been little more than scientific toys, but when our fuel oil is gone and our coal supply is low, they may come into their own.

Warmed by solar radiation, the earth itself acts as a radiator. Here again our fortunate atmosphere intervenes in the guise of a beneficent blanket. The earth being far cooler than the sun, its radiation is pitched at far longer wavelengths, all in the infra-red — mostly between 40,000 and 400,000 a. Most of this longer-wave, outgoing radiation is stopped by water vapour and carbon dioxide in the lower levels of the atmosphere, and much of what does escape these lower screens is absorbed by the high ozone screen mentioned above.

The earth's atmosphere in its complexity, therefore, acts like the glass of a greenhouse, letting in the sun-rays and keeping in the earth-rays. Were it not for this screening effect, the earth might be on the average more than 56° F. colder, and a far less livable place, than it actually is. Radiant solar energy received on earth is called 'insolation.' If the sun itself never varied, if its distance from the earth remained constant, and if one could measure insolation by holding a small surface perpendicular to the sun-rays for a stated time outside the ever-varying complexities of the earth's atmosphere, it would always be the same. But actually, at any time and place on the earth's surface, insolation is an ever-varying quantity. The south slope of a hill, for example, being more nearly perpendicular to the solar rays, receives more insolation in a day than a north slope, and is consequently regarded as a warmer and pleasanter place to live — at least in northern regions.

At any place on this spinning earth insolation mounts daily with the sunrise, reaches a maximum at some time during the day, and sinks daily with the sinking sun. Aside from the daily variation, the insolation received at different latitudes in any given part of the day of course varies greatly, owing to the slope of the earth's bulge, from the equator to the poles. And finally, this variation is increased with the rhythm of the seasons by the tipping of the earth's axis.

The natural enfeeblement of sun-rays spread thin, in arctic and sub-arctic regions, by the slope of the earth's surface, is further aggravated by the fact that these slanting rays must cross more atmosphere and suffer more absorption from it. Nor is this absorption uniform. The shorter wavelengths — ultra-violet, violet, and blue — are absorbed and scattered more; which scattering accounts for the blue colour of the clear daylit sky. The longer wavelengths — red and infra-red — are absorbed less and transmitted more; which transmission accounts for the reddish colors of sunset and sunrise.

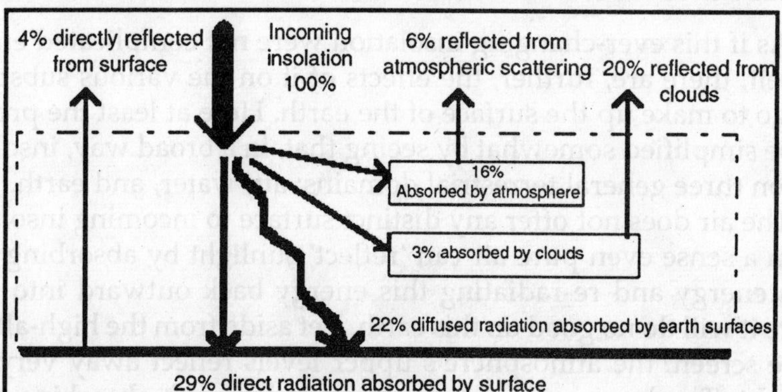

Fig. Insolation

Added to these continual but regular variations in the insolation received at any place on earth are the irregular changes imposed by an ever-varying atmosphere. Even when the sun is high overhead and the air ideally clear and pure, not more than four fifths of the incoming radiation reaches the ground.

On an average clear summer noon in temperate latitudes we do well to receive three fourths of the incoming insolation at sea level, though a larger proportion of course reaches earth in high mountain country. On a dark, cloudy day we are lucky to receive a fifth of the incoming energy even under a high sun, and must do with much less than this fraction if the sun is low.

Yet opposed to all these insolation changes are two factors tending to equalize them, at least for certain times and regions. When summer comes to either hemisphere, the days lengthen in its temperate zone, and near its pole there is no night.

However thin the arctic insolation may be spread by its obliquity, it lasts twenty-four hours of the day, while the equatorial day lasts only twelve hours. If there were no absorbing atmosphere, a place near the pole would actually receive more insolation during one of its summer days than a place near the equator; which is why far-northern summers are surprisingly warm, and stimulate the rapid growth of vegetable and animal life.

And when winter comes to the northern hemisphere, in which most of the world's people live, the eccentricity of the earth's orbit makes itself felt. As this orbit is no true circle, but an ellipse, the sun is about three million miles nearer the earth in our winter than in our summer. As a result its effective radiation in January is about seven per cent greater than in July, moderating the rigors of our chill winds.

TERRESTRIAL HEATING AND COOLING

As if this ever-changing insolation were not complicated enough in itself, there are, further, the effects of it on the various substances that go to make up the surface of the earth. Here at least the problem can be simplified somewhat by seeing that, in a broad way, insolation falls on three general terrestrial domains: air, water, and earth.

The air does not offer any distinct surface to incoming insolation. But in a sense even pure air can 'reflect' sunlight by absorbing some of its energy and re-radiating this energy back outward into space where it will do no good on this earth. Yet aside from the high-altitude ozone screen, the atmosphere's upper levels reflect away very little sunlight. The lower atmosphere, more laden with absorbing gases, dust, and haze, reflects and absorbs more of the incoming sunlight;

still it transmits most of it on to the earth's surface — that is, if the sky is reasonably clear. But if clouds intervene they reflect away about four fifths of the incoming energy. Water, somewhat like clouds, immediately reflects away much of the insolation that it receives. But in the case of water this reflected portion is not much more than a third, and the remainder is used in the thorough and beneficent fashion that makes a marine climate what it is. For one thing, the sunlight penetrates water to some depth, diffusing its warmth through a thick layer. Also, evaporation from the water surface, by its

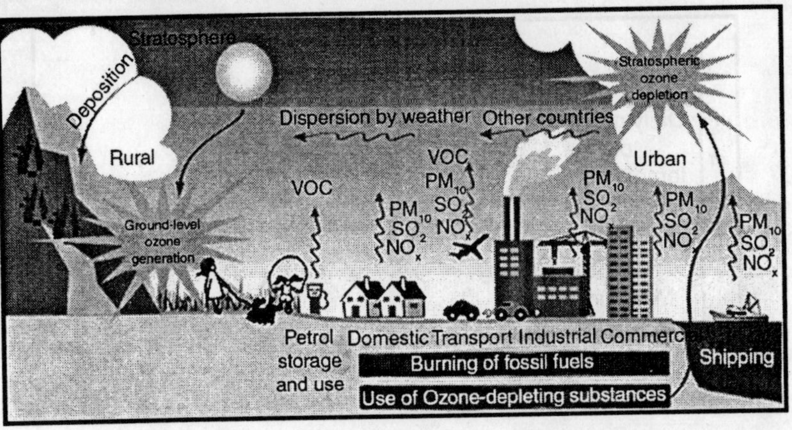

Fig. Effects of Sunlight on Air, Cloud, Water, Earth

Cooling effect, tends to check excessive temperature rises under a vertical sun. Finally, water itself is so constituted that a great deal of heat is needed to change its temperature much (it has a high 'specific heat,' in other words). And the addition or subtraction of a great deal more heat, even, is required to change its state — to freeze it, thaw it, vaporize it, or condense it. Water thus acts like a great, inert governor on temperature, and obstinately opposes any large temperature changes. Hence the main characteristic of a marine climate — a very small temperature change between day and night.

Land uses insolation very differently from water. Dry earth reflects away very little of the incoming energy. But it also transmits very little, so that daytime heating seldom extends below a depth of six feet, and even seasonal heating never gets below sixty feet. Practically all the incoming energy is absorbed in a narrow depth zone; and serves only to raise the temperature of the ground surface, which may be thirty or forty degrees hotter than the air a few feet above it. Gradually, though, this overheated earth warms the lower levels of air. That is why the summer temperature on Arizona deserts may go to 130° in the shade, why the whole Midwest occasionally simmers at over 100°.

Though the sun is highest and hottest near noon, it is some time before the outgoing earth radiation catches up with the incoming insolation (and until outgoing radiation balances incoming radiation, temperature must continue to increase). For this reason, maximum surface air temperature is usually recorded in the middle of the afternoon, three or four hours after the sun reaches its maximum intensity at noon. Forests have the effect of further delaying the daily maximum; and they also serve, in somewhat the same way as water, to moderate temperature extremes.

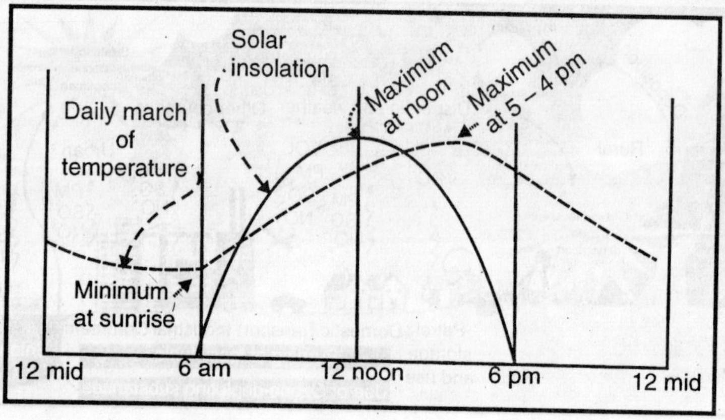

Fig. Hour-to-Hour March of Temperature Lags behind the Sun

At night, the heat in the earth's upper crust is quickly radiated upward into the atmosphere, and the ground temperature falls rapidly, cooling the air immediately above it. As this radiational cooling goes on most of the night, the lowest surface air temperature usually comes just before or at sunrise.

If clouds obscure the sky, they reflect some of this outgoing energy back to earth, and greatly moderate the nocturnal cooling process. But in any case the radiational cooling goes steadily on, at a faster or slower rate, during the dark hours. The only occasional stay to this night-time cooling is where great quantities of warmer air are being brought into an area by cyclonic winds.

During autumn in north temperate latitudes this night-time radiational cooling greatly exceeds, each day, the energy received from the sun. As a result, the earth gradually, from day to day, cools down to the low temperatures of midwinter, and does not begin to warm up again appreciably until late winter or early spring.

The annual march of normal temperature, like the temperature curve on any average day, lags somewhat behind the sun. Thus on land the coldest temperatures usually come late in January, while the warmest are to be expected late in July, lagging the sun by about a month. On the oceans, or along their leeward shores, the seasonal lag may be as much as two months.

The highest temperatures and the lowest are always encountered on land, and the characteristics of a continental climate are large temperature ranges between day and night, and larger temperature ranges between summer and winter. Whereas the annual temperature range is less than 5° F. over the equatorial oceans and less than 20° F. over all navigable oceans, it is more than 120° F. in the midst of Siberia.

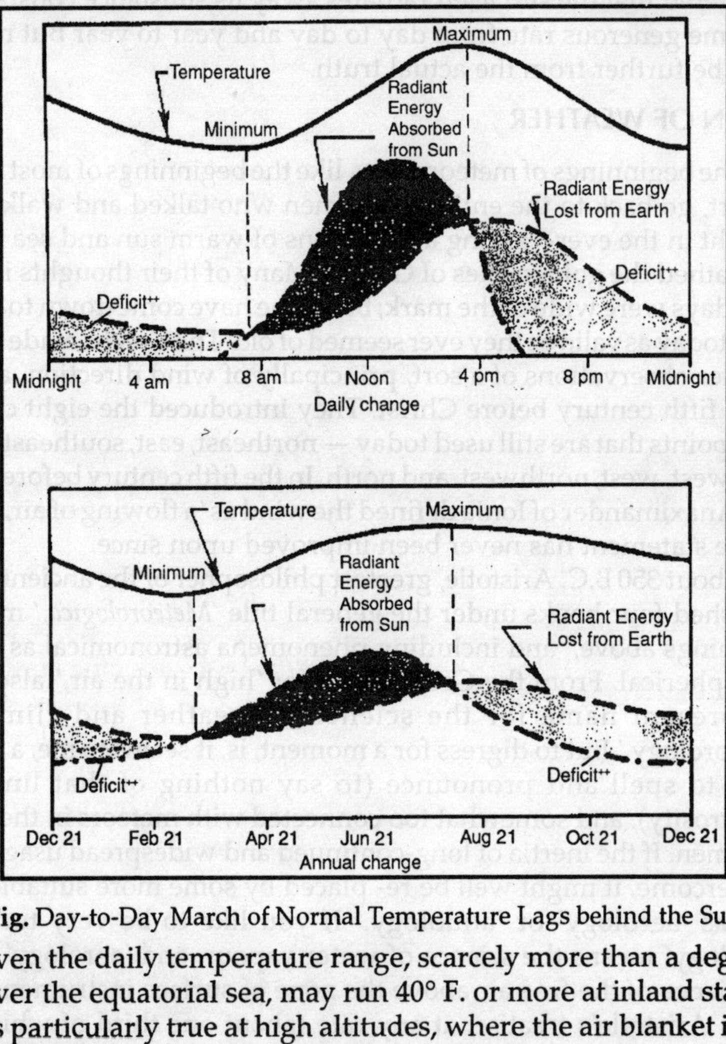

Fig. Day-to-Day March of Normal Temperature Lags behind the Sun

Even the daily temperature range, scarcely more than a degree or two over the equatorial sea, may run 40° F. or more at inland stations. This is particularly true at high altitudes, where the air blanket is thin; in parts of Tibet this daily range is 86° F. And on the moon, which has scarcely any atmosphere, it is perhaps 506° F.!

So far our account of solar radiation and earthly insolation has not touched on the first main result of earthly heating and cooling — that is to say, on the semi-permanent pressure centers and the winds of the world — or their tremendous modifying effects on earthly weather and climate. Nor have we considered at all the effects of moisture.

These things, and many others entering into the complex chain that links ultimate causes in the sun with ultimate effects in worldly weather, must be told of in their turn. And so far, also, we have considered that the sun itself radiates away its substance constantly at the same generous rate from day to day and year to year But nothing could be further from the actual truth.

BEGAN OF WEATHER

The beginnings of meteorology, like the beginnings of most science and art, go back to the enlightened men who talked and walked and thought in the ever-varying alternations of warm sun and sea change that bathed the golden isles of Greece. Many of their thoughts in those early days were wide of the mark; but some have come down to us, and stand today as valid as they ever seemed of old. The Greeks made regular weather observations of a sort, principally of wind direction, as early as the fifth century before Christ. They introduced the eight cardinal wind points that are still used today — northeast, east, southeast, south, southwest, west, northwest, and north. In the fifth century before Christ, also, Anaximander of Ionia defined the wind as 'a flowing of air,' which simple statement has never been improved upon since.

About 350 B.C. Aristotle, greatest philosopher of the ancient world, published four books under the general title *'Meteorologica,'* meaning 'the things above,' and including phenomena astronomical as well as atmospherical. From the Greek word for 'high in the air,' also, grew our present name for the science of weather and climate — 'meteorology.' But to digress for a moment, is, it seems to me, a clumsy word to spell and pronounce (to say nothing of that linguistic monstrosity), and somewhat too connected with meteors in the minds of laymen. If the inertia of long-continued and widespread usage could be overcome, it might well be re- placed by some more suitable term, such as 'aerology' or 'atmology.' If you like to be very technical, 'atmology' means the science of water vapour, and 'aerology' means the science of the free air above the zone of surface instruments.'

In Aristotle's treatise, at any rate (about one third of which was devoted to the science of the atmosphere), many weather questions were considered — the formation of fog and clouds, dew and frost, rain and snow and hail; theories of winds, thunderstorms, and whirlwinds The whole work was marked more by powerful imagination than by scientific method. Yet it was the earliest comprehensive work on meteorology, and remained for fifteen centuries the best.

One of Aristotle's disciples in the famous peripatetic (or 'walking-about') school of philosophy at Athens was a man whom Aristotle nicknamed Theophrastus in recognition of his graceful conversation. Like most early philosophers, Theophrastus ranged over a broad field of subjects — psychology, botany, physics. But he also found time to write a treatise on wind and weather which included some weather proverbs not unlike those that find credence today.

In ancient times life was leisurely and new developments came slowly. Rain, always vital to husbandry and farming, was the first weather element to be accurately measured. (Such measurement requires, indeed, nothing more complicated than an open bucket and a ruler.) As early as the first century A.D. rainfall measurements were made in Palestine, and the resulting rainfall records were preserved among the Jewish religious writings.

Much later, by 1450, rain gauges were in use by the Koreans of China. But in the Middle Ages the frontiers of aerological knowledge were already widening in another direction. During the eleventh century some Arabian astronomers, watching sunrise and sunset in the clear air of the Near-Eastern desert, made the first rough estimate of the height of the upper atmosphere. They deduced from the duration of twilight that the earth's atmosphere extends in appreciable density to heights around fifty miles, and later experiments have confirmed their findings.

Simple facts and principles such as those mentioned above are easy enough to understand. And such simple principles are really the fundamentals of all human knowledge. The hard part of any science, for an unaccustomed layman, is its symbolism and technical jargon. Talk about 'lapse rate,' for example, sounds very mysterious until you understand that this term is merely aerologists' slang for the way temperature decreases with increasing height in the atmosphere.

Perhaps the easiest approach for a present-day layman to the complex and comprehensive science of modern aerology is through the story of how weather science developed in an era when all science was simpler and briefer than it is today.

The Invention of Instruments

If meteorology may be said to have dawned in ancient Greece, its sun really rose in the sixteenth-century Tuscany of the Medicis. For weather work began to assume some dignity as an independent science when quantitative measurements of its several elements became possible; that is, when accurate weathermeasuring instruments appeared. The first of these (aside from the rain gauge), and one of the most important, was the thermometer. Back in the third century Philo of Byzantium and Hero of Alexandria had devised some primitive thermoscopes.

The instruments themselves had probably no great value; but one of Hero's writings came, centuries later, to the attention of the first great experimental scientist — Galileo Galilei. He and his associates were to make northern Italy the cradle of instrumental meteorology and instrumental physics.

In 1592, Galileo, then at the University of Padua, constructed the first practical thermometer. Bearing small resemblance to the modern instrument, it consisted of a large bulb filled with air which, by its expansion when heated, forced downward the water level in a glass tube. Somewhat later, about 1612, Galileo is thought to have devised, in essence, the modern thermometer — a small glass bulb of liquid free to expand into a sealed capillary tube. (The liquid was alcohol, which is still used in many modern instruments.) Thermometers of this type, at any rate, were used around 1640 by the gentle and courageous Ferdinand II of Tuscany, one of the most enlightened of the later Medicis.

Around 1720, Gabriel Daniel Fahrenheit, a Danziger settled in Holland, started using mercury as a thermometer liquid. Fahrenheit also devised the first practical and convenient thermometer scale — the Fahrenheit scale still in ordinary use among all the English-speaking peoples. And about 1742, in Sweden, the astronomer Anders Celsius devised the Centigrade thermometer scale which, turned right side up, is used in all scientific work throughout the world today.

Meanwhile the barometer had appeared. Before 1640, the same Grand Duke Ferdinand II of Tuscany had further stimulated the march of knowledge by ordering that a deep well be dug near Florence. Such an ordinary action might at first glance have been thought free from future scientific implications; yet it was found, to the surprise of all, that water would not rise in the well pipe, even when the air was drawn out of it, to a height greater than about thirty-two feet. Galileo, consulted near the end of his life's labors, judged that 'nature's abhorrence of a vacuum,' until then thought absolute, appeared to have limits. Before his death he suggested that liquids of different densities should rise to different heights in an evacuated tube, and passed the problem on to his pupil, Evangelista Torricelli.

About 1643, Torricelli put the experiment on a more convenient scale by adopting mercury, the heaviest of all liquids, which rose only thirty inches in an evacuated tube. He then concluded that, in the words of Millikan and Gale, 'the rise of liquids in exhausted tubes is due to an outside pressure exerted by the atmosphere on the surface of the liquid, and not to any mysterious sucking power created by the vacuum.' Torricelli proved this by placing his whole barometer in a bell jar and pumping out the air, whereupon the mercury column sank in the barometer tube. Torricelli's barometer, unchanged in general principle though considerably refined in form, is used for all accurate pressure measurements today.

Barometry now moved by leaps and bounds. In France, an impoverished but intelligent aristocrat named Blaise Pascal reasoned that air pressure, like pressure in other fluids, should diminish as one goes higher in the atmosphere. Lacking a hill or which to try the experiment, he wrote his brother-in-law Perrier in the south of France. In 1648, Perrier carried a mercury barometer up a sizeable mountain (Puy-de-Dôme), and was 'ravished with admiration and astonishment' when he found the mercury column sinking in the tube.

The surveying aneroid and the aeronautic altimeter were now possible in principle, and became a reality with the invention of small, easily portable aneroid barometers in 1844. The barometer was also used, even in the seventeenth century, to indicate daily small changes in atmospheric pressure, which before long came to be associated with weather changes; high pressure vaguely foretelling fair weather, and low pressure, foul. But the barometer's reputation as a 'weather glass' has been, and still is, over-inflated.

The barometer and thermometer soon became, and still remain, the most important instruments in aerology, but other developments followed fast. In 1653 that same Grand Duke Ferdinand of Tuscany, who had used the first accurate thermometers and whose well had begun the whole barometer sequence, made another outstanding contribution to science by establishing several weather-observing stations throughout northern Italy -the first weather network ever projected and the germ of modern national weather-reporting services. Ferdinand sought also to establish an international chain of weather stations, but this plan suffered the usual fate of ideas broached before their time.

Around 1667 an Englishman named Robert Hooke devised the

earliest type of anemometer, giving for the first time a numerical measure of wind strength. This simple device was merely a rectangular plate hung by its top so that increasing wind pushed it out to an increasing angle from the vertical.

Rotating anemometers came into use by about 1724, and the instrument was further improved some fifty years later by the Swiss geologist, meteorologist, and alpinist, Horace Benedict de Saussure. Saussure also pioneered much of the early work in moisture measurement; and he invented the hair hygrometer, which indicates atmospheric moisture in simple and direct fashion by virtue of the fact that a strand of human hair expands when it is moist, contracts when it is dry.

Beginnings of Analytic Thought

The great trade winds of the tropics had been known and used for centuries before anyone sought rationally to explain them in the light of known facts. In 1686, Edmund Halley, British Astronomer-Royal, soon after calculating the orbit and predicting the return of his famous seventy-seven-year comet, published a paper entitled *'An Historical Account of Trade Winds and Monsoons Observable in the Seas Between and Near the Troppicks, with an Attempt to Assign the Physical Causes of the Said Winds.'*

Halley rightly concluded that monsoons -winds blowing alternately from sea to shore and shore to sea -were caused by the temperature differences between land and water. He also demonstrated that the sun's heat near the equator was the primary cause of the trade winds. Halley, incidentally, was the first astronomer to chart the stars of the southern hemisphere, and spent two years at sea studying the magnetism of the earth.

But Halley overlooked one very important factor in the winds of the world, which another Englishman, George Hadley, discovered in 1735. This factor is the deflective force due to the earth's rotation, which in the northern hemisphere causes the southward-drifting trades (or any other winds from the north) to turn westward, and leads the northward air drifts in temperate latitudes (or, in fact, any winds from the south) to turn eastward.

After 1800 there began in all scientific fields that modern multiplication of research effort which subordinates individual names and dates in one large, long, and accelerating advance. But some few names do stand out, nevertheless, in nineteenthcentury meteorology. For example, an American weather student named James Pollard Espy published in 1841 a book called *'The Philosophy of Storms.'* Espy was the first scientist to point out the convection (or large-scale rising of air) that takes place in areas of low pressure or cyclones.

Around 1850, Heinrich Wilhelm Dove, director of the Royal Meteorological Institute at Berlin and one of the greatest of all meteorologists and climatologists, brought out in Germany his celebrated book on the *'Law of Storms.'*

He tried to give a fuller explanation of the great wind systems of the world, postulating two main circulations in each hemisphere: one between the equator and the tropics, the other between the tropics and the pole. He studied keenly the origin of large-scale storms -the cyclones or low-pressure areas of temperate latitudes — and concluded that they were due to the continual strife between tropical air currents flowing up from the south and polar air currents flowing down from the north..

Around 1850, too, Matthew Fontaine Maury of the United States Navy made his famous researches into the prevailing winds of the seven seas. Maury finally prepared complete charts of these winds, including the tropical easterly trades and the temperate-zone westerlies. By using his charts, sailing-ship captains found that they could cut the time of long voyages by twenty-five per cent — say from four months to three on the England-Australia run. Similar average wind-and-current charts are used by all steamship mariners today.

After Dove and Maury came many distinguished men, advancing theoretical meteorology in various particulars. William C. Redfield in America studied West Indian hurricanes and the storms of the Atlantic coast, finding that 'a cyclone is constituted by a considerable mass of air endowed with a rapid movement of rotation in the direction opposite to the hands of a watch.' William Ferrel, also an American, worked for many years on the problem of the world's air circulation, and published theoretical solutions that were accepted for some time.

At the time of the American Civil War, Sir Francis Galton, working in England, wrote: 'I have deduced . . . the existence not only of cyclones, but of what I dare to call anti-cyclones.' And today, the anti-cyclone, or high-pressure area (which represents the controlling centre of an air mass), is beginning to assume even more importance in modern weather forecasting than the formerly worshiped cyclone, or 'low.' In Holland, Buys Ballot framed his oft-reiterated law of wind and pressure.

'If you stand with your back to the wind in the northern hemisphere, then the low pressure will be on your left hand'; and developed other important pressure-wind relations. Towards the turn of the century, Hugo Hildebrand Hildebrandssen of Sweden and Leon

Teisserenc de Bort of France revised Ferrel theory of world winds, and in their *'Fundamentals of Dynamic Meteorology'* reduced all previous ideas of the planetary circulation to something like the complete and unified theory that is generally accepted today.

WEATHER-REPORTING AT NEAR THE SPEED OF LIGHT

Nearly two hundred years had come and gone before the Grand Duke Ferdinand's idea of a network of weather-reporting stations was again seriously tried in Europe. In the opening years of the nineteenth century, the Chevalier de Lamark, working in France with Laplace, Lavoisier, and others, had established a system of weather stations and published a series of yearly weather summaries. And in 1820, Heinrich Wilhelm Brandes, a German, had conceived a simple yet profoundly important idea that the widespread phenomena of the weather could best be studied by plotting on a map various weather observations taken at the same time, covering as large an area as possible.

(The weather data used in Brandes's maps, though, was thirty-seven years old.) In 1821, the American, Redfield, had also drawn these synoptic maps of American weather conditions. But in those days, of course, the synoptic maps could not be made until slow mails had arrived from distant places — until the weather being studied had come, done its worst, and gone.

In 1844, the first commercial telegraph clicked from Baltimore to Washington the devout message: 'What hath God wrought!' At the same time men were suggesting in various parts of the world that the new electrical marvel should be used to send weather reports from outlying stations into central weather offices. But the first actual telegraphic weather reports were sent in 1849 at the suggestion of the great American electro-physicist, Joseph Henry.

He and Espy organized the first regular weather service in the United States. Like many another admirable American agency of peace, this weather service was born and nurtured in our War Department; and the first official American weather forecasters were Army officers detailed to that unac customed duty. By 1891 the meteorological section of the Army Signal Service became the Weather Bureau, which has now grown into a large, ubiquitous, and immensely useful organization.

Espy and Elias Loomis greatly improved weather-plotting methods. Under them the daily weather map grew towards the romantic tapestry it is today, showing at a glance all the morning's weather over the whole of North America from Alaska to Panama, or even over the whole northern hemisphere.

Daily weather maps were introduced into England by Admiral Fitzroy, who in 1861 hazarded a storm forecast that actually came true. Soon after, in France, the astronomer Le Verrier began publishing a daily, up-to-the-minute weather map. Meanwhile Fitzroy, together with Siljeström and Lilliehöök in Sweden, confirmed and elaborated some of the storm theories of Dove.

The Blue Depths Above

Long before kites, balloons, and airplanes appeared in human history, there were the mountains, already fashioned by geologic forces to the weather uses of man. Yet aside from the religious mountaineering of the Old Testament, history records few early mountain climbs of importance. One of the most notable occurred about a hundred years after the birth of Christ, when the just and courageous Roman Emperor Trajan climbed the two-mile peak of Mount Etna to watch a Sicilian sunrise.

High mountaineering then apparently lapsed until the Middle Ages, when several more-than-two-mile summits in northwestern Europe were climbed. It was not until the eighteenth century, though, that aspiring mountaineers began to surmount three-mile summits covered with eternal snow, carrying with them thermometers and other instruments to measure the meteorological entities at these cold and tempestuous heights.

In the eighteenth century, also, meteorological soundings into the free air made a modest beginning. A Scotsman of Glasgow named Alexander Wilson measured atmospheric temperatures up to half a mile or so one summer day in 1749, by sending aloft thermometers attached to several kites tied in tandem to one long string. Lifted by a strong and steady breeze, the high- est kites disappeared among woolly-white cumulus clouds.

But at intervals, a burning fuse along the upper kite string dropped the thermometers (which were of a slow-responding type), and being tied to bushy tassels of paper, they parachuted quickly but safely to earth. Doctor Wilson and his assistant Melvill ran immediately to recover them, and thus obtained approximate indications of the existing temperatures aloft.

In 1752, Benjamin Franklin in the American Colonies made his famous discovery of the nature of lightning by flying a kite into the teeth of a thunderstorm. Luck had Franklin by the hand on that auspicious day, for if the side-flash that traveled down his wet kite string and sparked across to his knuckle had been a full-sized bolt, his career would have ended then and there.

Following after Franklin, other early kite-flying experiments served to show some interesting facts about the directions and patterns of lower-level winds. But a better idea for upper-wind indication was forthcoming when, in 1809, Thomas Foster in England began to use small balloons, drifting freely upward and away, for this purpose. Thus began pilot balloons, which are now released by aerologists all over the world at the rate of hundreds each day.

During the late eighteenth and early nineteenth centuries the free balloon or 'aerostat' was in its heyday, daily carrying men to various adventures and not seldom to disaster. The first really accurate temperature measurements of the upper air were made by the Franco-American aeronauts, Jeffries and Blanchard, in 1784.

Soon the man-carrying balloon was in common use for atmospheric soundings. In the eighteen-fifties John Welsh of England, with Charles Green as his aeronaut, made several ascents up to an altitude of about twenty thousand feet, observing pressure, temperature, and humidity at all levels. Welsh was actually reaching in his old-fashioned aerostat higher altitudes than those attained by the scores of sounding airplanes that were to rise into American skies eighty years later.

Among the Britons who watched Welsh's ascents with telescopes was a Scottish assistant at Greenwich Observatory named James Glaisher. With Henry Coxwell as aeronaut, and a great sixty-foot balloon capable of lifting more than two tons, Glaisher later made twenty-eight high-altitude balloon soundings Glaisher and Coxwell also, incidentally, took the first aerial photographs and dropped the first aerial bombs.

On one ascent these mid-Victorian aeronauts set a world's altitude record that stood for many years, and very nearly cost them their lives in the bargain. It was the afternoon of September 5, 1862. The ponderous aerostat left the ground near London at 1.03 P.M., the surface temperature being 59° F. and the sealevel barometer standing at about 30 inches of mercury. Ten minutes later the balloon passed through a cloud layer extending from six to seven thousand feet altitude, in which the temperature was only 36°. As Glaisher wrote in the book he later published:

> . . . Momentarily the clouds became lighter, and on emerging from them . . . a flood of strong sunlight burst upon us, with a beautiful blue sky without a cloud, and beneath us lay a magnificent sea of clouds, its surface varied with endless hills, hillocks, and mountain chains, and with many snow-white tufts rising from it. I here attempted to take a view with a camera, but we were rising with too great rapidity

and revolving too rapidly to enable me to succeed. We reached the height of two miles at 1 h. 22 m., where the sky was a dark blue, and from whence the earth was visible in occasional patches beneath the clouds. . . .

They reached three-mile altitude at 1.28 P.M., climbing at a better rate than many modern airplanes could maintain. As the balloon continued its rapid ascent, their barometers continued to fall, and the temperature dropped to freezing, then to zero, and then to -2° F. as they reached an altitude of five miles. Continues Glaisher:

. . . Up to this time I had taken observations in comfort, and experienced no difficulty in breathing, whilst Mr. Coxwell, in consequence of the exertions he had to make, had breathed with difficulty for some time. Having discharged sand, we ascended still higher; the aspirator became troublesome to work; and I also found difficulty in seeing clearly. At 1 h. 51 m. the barometer read 10.8 in. About 1 h. 52 m. or later, I read the dry-bulb thermometer as minus 5°; after this I could not see the column of mercury in the wet-bulb thermometer, nor the hands of the watch, nor the fine divisions of any instrument.

I asked Mr. Coxwell to help me read the instruments. In consequence, however, of the rotary motion of the balloon, which had continued without ceasing since leaving the earth, the valve-line had become entangled, and he had to leave the car and mount into the rigging to readjust it. I then looked at the barometer, and found its reading to be 9¾ in., still decreasing fast, and implying a height of 29,000 feet.

After Coxwell had climbed into the balloon's rigging, Glaisher lost consciousness, and collapsed against the side of the basket with his head on the edge. Coxwell, struggling heroically on the waist-like upper ring with the fouled valve-line and the slippery frost that was forming on balloon and rigging, soon froze both his hands. He felt insensibility coming over him. But with a final effort he managed to seize the valve-line with his teeth, and dipped his head several times. The balloon stopped rising, paused, and at last began to descend.

Coxwell managed to place his useless arms on the ring, whence he dropped down into the basket beside Glaisher, nearly knocking the bottom out of it. As the descent continued, Glaisher revived, and poured brandy over Coxwell's blackened hands. The aerostat continued downward, more and more rapidly. Enough ballast remained, though, to prevent a crash landing. At about 2.30 P.M., after only an hour and a half aloft, the aeronauts landed in a large pasture near Ludlow.

Towards the end of the nineteenth century self-recording instruments were developed which made it unnecessary for men to go aloft after aerological data, provided there was no hurry about getting the data back. The smaller sounding balloons, or *ballon-sondes*, could go far higher without human danger or discomfort — easily to altitudes around ten miles and perhaps with good luck to twenty miles or so. Teisserenc de Bort pioneered these high-altitude, small-balloon soundings in France and elsewhere.

In 1901, he discovered the stratosphere or isothermal layer — that still, frigid region where temperature stops decreasing with increasing height. De Bort sent up his balloons from various points in Europe, and from his steam yacht in the South Atlantic Ocean. In Germany, Richard Assmann also sounded into the stratosphere with self-recording instruments. In the United States Abbot Lawrence Rotch sent up many sounding balloons to great heights, and collaborated with de Bort in studies which showed the variation in height of the stratosphere base from pole to equator.

PHYSICAL WEATHER ANALYSIS

Aside from the discovery of the stratosphere and a better understanding of the wind-and-climate belts of the world, the half-century from 1865 to 1915 saw few fundamental advances in weather analysis and forecasting, though towards the end of this period the development of seagoing radio greatly broadened the areas that could be covered by weather reports and maps.

Until the time of the First World War meteorologists everywhere were accustomed to draw nicely rounded low-pressure areas and highpressure areas on their weather maps, and to forecast weather more empirically than rationally. The burden of this empirical thought was mostly concerned with what in the way of average weather usually occurred in certain geographical portions of 'lows' and 'highs,' and with where these lows and highs on the average were most likely to move — in general, more with questions of average 'what' and 'where' than with questions of specific 'how' and 'why.'

But around 1917 it was discovered that most temperate-zone cyclonic lows, even when viewed in two dimensions, are not nicely rounded at all — that they are, in fact, usually cut up into pie-like sectors of diverse weather by pronounced lines of wind shift and discontinuity called 'fronts.'

And it was found also that the anti-cyclonic highs which seem to drive these cyclones and fronts are usually associated in rather complex fashion with the centers of great masses of air, several miles thick and several hundred miles wide. These 'air masses' may be relatively warm or cold, moist of dry; and they may stagnate for a time in one region, and then perhaps travel far over land or water while keeping to some extent the climatic characteristics (such as temperature and humidity) acquired in their last place of sojournment.

This new way of looking at weather phenomena originated in Scandinavia as one beneficent result of a stern wartime necessity. From 1914 to 1918, Norwegian aerologists were cut off from outside weather reports and broadcasts, which the belligerent nations naturally suppressed because of their possible usefulness to enemy raiders. Turned thus inward by force of circumstances, Bjerknes and others of the Bergen school began to draw large-scale regional weather maps, studying their own regional weather with minute thoroughness and applying to this study the physical principles that govern the behaviour of all fluids. The result was physical weather analysis or 'airmass-and-front analysis.' During the nineteen-twenties and nineteen-thirties it revolutionized the whole theory and practice of temperate-zone weather forecasting.

Significant and revolutionary as these recent developments in aerology are, they have disproved no *physical* principles that came before them. Nor will they. The wind is still 'a flowing of air,' as the ancient Ionian defined it. The rise of liquids in exhausted tubes is still due to an 'outside pressure exerted by the atmosphere on the surface of the liquid,' as Torricelli stated during the Renaissance. The large-scale storms of the temperate zones are still due, as Dove discovered when the United States was young, to the continual strife (now garbed in the nomenclature of air masses and fronts) between tropical and polar air currents. There is nothing absolutely new under the sun. But the world does move, and aerology moves with the forward march of science.

Chapter 4

Surface Winds

The winds of the world have been blessed, cursed, and generally noted by men since the dawn of human thought and wonder. They have long been of vital importance to mariners, and even a modern *Queen Elizabeth* burns up hundreds of extra dollars in fuel oil while bucking a winter gale. Nor, in the air ocean, does any fast modern air liner fly before its captain has taken full account of the upper winds to be expected along its course — winds that may not be able to change a super-powered schedule, but will certainly control the amount of gasoline fed to the engines.

Two remarkable things about the whole broad pattern of winds in the air ocean above us are: first its complexity, and second its variability. Along the Atlantic seaboard, for example, if the surface wind sets in from easterly and clouds gradually lower, we rightly expect a storm; yet careful sky-watching will reveal that all the upper storm clouds are approaching, not from the murky east, but from the distant west. It is no uncommon thing, either, to see anywhere in temperate latitudes a lower deck of scattered or broken clouds drifting up from the southward, while far above a higher and thinner cloud layer at the same time drifts noticeably from west to east. In clear weather an ascending old-fashioned fire balloon (such as those which used to grace Fourth of July evening skies, and set fire to more than a few innocent housetops, of a generation ago) may drift first westward or northward as it rises from the ground, and then, at considerable altitude, seem to come back, cross its earlier track, and finally disappear to eastward.

Even a child's kitestring in March may form a pronounced bend that points unmistakably to differences between winds on the earth's surface and those even slightly above it.

STUDYING THE SURFACE WINDS

Wind Direction

The ancient Greeks distinguished between eight wind directions in all, and the same classification is used in many of our weather observations today: north, northeast, east, southeast, south, southwest, west, and northwest. But in many modern wind reports, such as those along the airways, sixteen compass points are used: north, north-northeast, northeast, east-northeast, east, east-southeast, and so on. The 'direction' of a wind, as used in weather reports, is always the *direction from which it blows*.

There are simple ways to note wind direction — by the leaves and branches of trees, by the waves on water. In light airs a pinch of tossed dust, or a puff of smoke, drifts with the breeze. (The 'wet-finger' method has been much overrated.) The simplest way of all, and one of the most effective, is to face into the wind, turning slowly from side to side. If you are off centre,

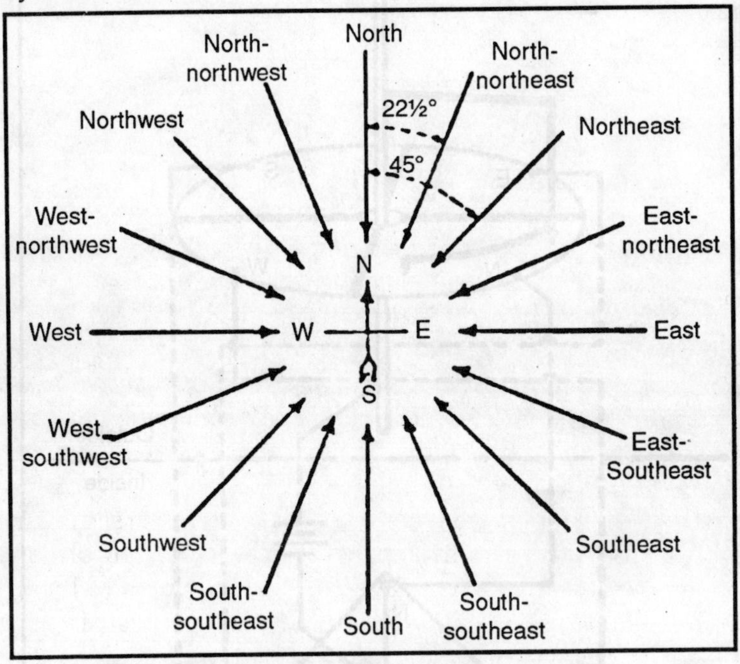

Fig. The Sixteen Winds of Modern Aerology

you can easily feel the greater wind pressure or the greater cooling effect on one side or other of the face. In unknown terrain, the direction can then be found from a compass or from the sun or stars.

Wind Vanes

The instrument universally used for telling wind direction is the wind vane. All over the world it appears in many forms, but its essentials are always the same. The vane part itself is usually made in the form of a balanced arrow free to swing about a vertical spindle. The long, wide tail of the arrow is deflected by the slightest side draft, and keeps the point directed into the wind. Beneath the turning arrow two fixed cross-arms are usually mounted at right angles; and their ends, which should be accurately aligned on the four cardinal points of the compass, are lettered N, E, S, and W. To read a wind vane of this simple type, of course, one has to go outside and note the heading of the vane.

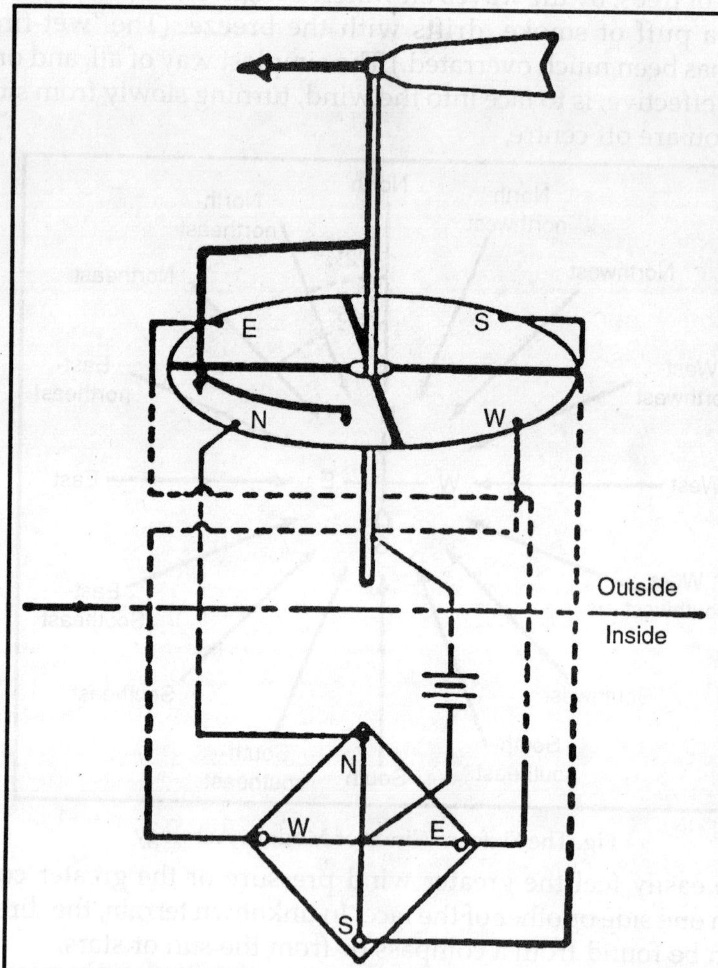

Fig. Principle of a RemoteIndicating electric Wind Vane

In weather observatories various kinds of remote-indicating and recording wind vanes are used. The simplest of these in principle, though not very convenient to build, is a long wind vane spindle that extends right down through the roof of the building and connects with either an indicating pointer above a compass scale or a pen that circles about a recording drum which can be moved slowly up or down by clockwork. A more flexible and convenient type of remote-indicating wind vane, used at most weather-observing stations today, operates by electricity.

The circle below the vane is divided into four metal segments, and metal contactors attached to the vane roll over these segments in such a way that current is transmitted to the indicator in the observing station by appropriate direction wires, lighting small electric bulbs labeled N, E, S, and W. The same principle is used in making continuous records of wind direction on a slowly revolving drum; here electric magnets operate pens periodically so as to record dots in the N, E, S, or W sections of the chart. Such indicators and recorders cover eight wind directions, for if the wind is northwest, say, both the N and W bulbs light. The most modern wind-direction indicator operates by means of a device that is gradually coming into all sorts of uses in our modern mechanical world — the self-synchronous electric motor or 'selsyn.' For measuring wind direction, the wind vane turns one selsyn, and the other selsyn turns either an indicating hand above a compass scale or a recording pen around a drum. Aside from wind indications, the selsyn principle is also beginning to be used in various other remote-indicating weather instruments.

Wind Speed

Long before the invention of instruments that would accurately measure the wind's speed, people used to guess at it. Lacking instruments today, they still do. Every smallship sailorman has to be constantly gauging wind force by the actions of his boat and the look of the water; and on land, also, one can observe many indications of wind force. Of all the various estimating scales that have appeared at various times in the world, the scale originally devised about 1805 by Admiral Beaufort, of the British Navy, has been used longest and most widely. Another wind classification, and a simpler one, is that generally used in weather forecasting:

Calm	0-5 m.p.h.
Light	5-15 m.p.h.
Moderate	15-25 m.p.h.
Fresh	25-35 m.p.h.
Strong	35.50 m.p.h.
Gale	Over 50 m.p.h.

Estimating wind velocity is very useful in its way, and does well enough if no instrumental measurements are available. But the plain truth is that judging wind speed involves a great deal of guesswork, unless one has had considerable practice. In most serious aerological work instrumental wind-speed measurements are needed, and even these often leave much to be desired. One ordinary wind indicator, however, comes near being a wind-measuring instrument — a flag.

Table. Wind Speeds

(Based on the twelve-point Beaufort scale)

Force	Description	Wind Speed (in statue or land miles per hr., approx.)	Wind Speed (in meters per sec., approx)	Wind Pressure (in pounds per sq. ft., approx.)	Effect on Ordinary things
0	Calm	0	0	0	Smoke rises straight up.
1	Light air	2	.9	.01	Smoke drifts at strolling speed.
2	Light breeze	5	2.2	.06	Smoke drifts at dog-trot speed; leaves rustle; wind felt on face.
3	Gentle breeze	10	4.5	.26	Leaves and twigs move.
4	Moderate breeze	15	6.7	.58	Small branches move; dust and loose paper raised.
5	Fresh breeze	21	9.4	1.1	Small trees sway.
6	Moderate wind	28	12	1.9	Large branches in motion; wind begins to whistle.
7	Strong wind	35	16	3.1	Large trees in motion; wind difficult to walk against.
8	Gale	42	19	4.4	Tree twigs break; walking progress slow.
9	Strong gale	50	22	6.4	Slight structural damage.
10	Whole gale	59	26	9.1	Considerable structural damage; exposed trees up-rooted.
11	Storm	68	30	12	Widespread damage; sel-dom experienced on land.
12	Hurricane	Above 75	Above 34	Above 15	Widespread destruction.

Of course flags vary greatly in size and weight, and their indications vary correspondingly. But ordinary flags enable one to guess more closely at the wind, if its speed is moderate, than the motions of trees and the like. Up to about thirty land miles an hour, increasing wind speed makes a flag stream out further from its staff - alongside which it hangs in a flat calm. In a ten-mile breeze, most flags fly about one third of the way between the staff and the horizontal. At twenty miles an hour, they are something like two thirds of the way up to the horizontal, or a little higher. At thirty miles an hour or more, they stream out along the horizontal or slightly above it.

Anemometers

Similar to the flag in principle is the simplest wind-speed instrument, known as the 'pressure-plate anemometer.' It is merely a weighted board or a metal plate, hinged along its upper edge so that it can swing out at various angles in response to various wind speeds. Along the lower edge of the plate is a stationary scale that can be calibrated to read directly in miles per hour or Beaufort force units.

In portable form the whole device can be held in the hands; or at a fixed location it can be permanently mounted, free to turn on a vertical spindle along with a wind vane that keeps it facing into the wind. This pressure-plate anemometer has one serious disadvantage, though — in gusty winds it tends to swing like a pendulum, giving faulty indications.

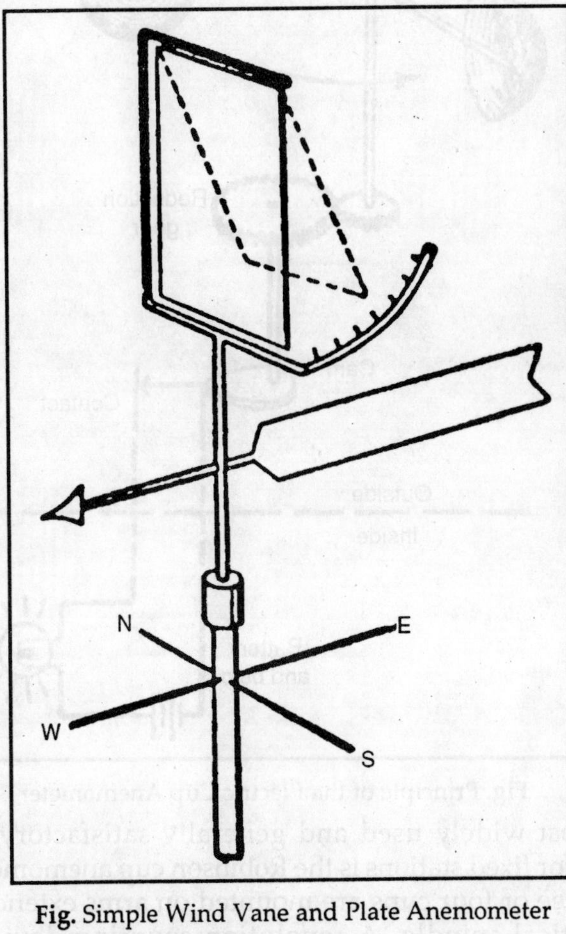

Fig. Simple Wind Vane and Plate Anemometer

Another type of wind-indicating instrument is the Pitot tube widely used in aviation. A small open tube, kept pointed into the wind by a wind vane, is connected by airtight tubing to some sort of pressure-indicating device. This latter may take the form of a mercury-filled glass U-tube or J-tube manometer, an aneroid vacuum chamber, or a system of liquid chambers and floats which can be connected to a recording pen. The Pitot tube responds very quickly to any change in wind speed, and is hence useful for indicating and recording the gustiness, or changeability, of the wind.

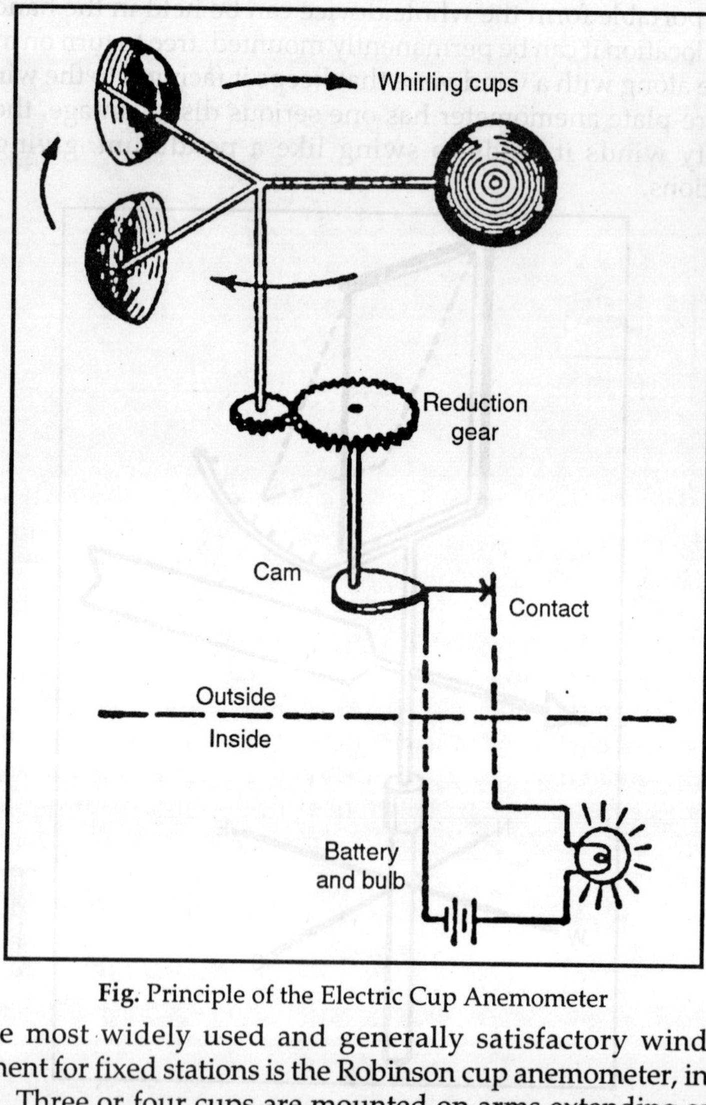

Fig. Principle of the Electric Cup Anemometer

The most widely used and generally satisfactory wind-speed instrument for fixed stations is the Robinson cup anemometer, invented in 1846. Three or four cups are mounted on arms extending outward from a vertical spindle. A revolution-counting device is usually arranged to close an electric contact for every sixtieth of a mile (or other convenient unit) of wind that passes the station. The electric contact flashes a light, turns on a buzzer, or operates a recording pen on a revolving drum. To get the actual wind speed with this instrument, of course, a simple computation is necessary; usually you count the light flashes in one minute, which gives the wind speed in miles per hour.

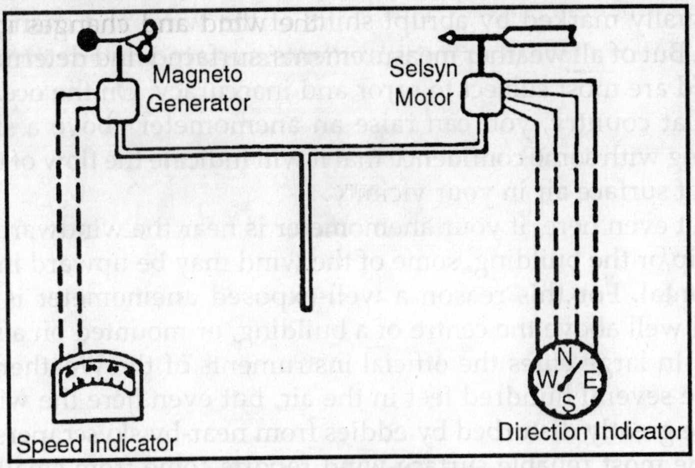

Fig. Electric Wind-Measuring Instruments

A newer and more responsive instrument of the cup-anemometer type indicates wind speed continuously on a dial. Equipped with light-weight cups, it responds within half a second to varying gusts. The exposed cups, whirling in the wind, drive a small electric magneto-generator that is connected by two wires to a distant voltmeter, which is calibrated to read directly in miles per hour of wind speed.

This magneto-generator instrument, used in conjunction with a four-segment electric wind vane, is probably the best all-around wind-measuring instrument available today. It is a revealing and fascinating study in the amazing variability of surface-wind patterns to sit in a large weather station and watch the dial of a magneto-generator anemometer during a squall or thunderstorm, or even during a strong wind that seems steady to the senses alone.

The flickering needle of the voltmeter clearly shows the existing gustiness by jumping, say, from twenty miles per hour up to thirty, to forty, back to twenty, up to fifty, back to thirty, and so on — all in the space of a few minutes.

It is easy to see why an airplane landing too slowly in a gusty wind is sometimes suddenly bereft of 'lift,' lack of which instantly results in a probably fatal crash — for once your flying speed (relative to the air) falls below a certain critical minimum (the stalling speed), your airplane will 'fly' like the proverbial stone.

Wind Measurements

Wind-direction and velocity measurements are very important in modern weather technique, as the fronts between unlike air masses are usually marked by abrupt shifts of the wind and changes in its speed. But of all weather measurements, surface-wind determinations on land are most subject to error and inaccuracy. On the ocean, or in very flat country, you can raise an anemometer above a ship or a building with some confidence that it will indicate the flow of the main body of surface air in your vicinity.

But even here, if your anemometer is near the windward wall of the ship or the building, some of the wind may be upward instead of horizontal. For this reason a well-exposed anemometer is usually placed well above the centre of a building, or mounted on a separate tower. In large cities the official instruments of the Weather Bureau may be several hundred feet in the air, but even here the wind flow may be greatly disturbed by eddies from near-by skyscrapers.

The most reliable surface-wind reports come from small islands in the ocean, or from stations surrounded by absolutely flat country. In hilly or broken country, or in deep valleys, the measured surface wind may bear not the slightest resemblance to the general, regional drift of lower air — it may be much slower, or even somewhat faster, and it may come from almost any direction, all depending on the manifold peculiarities of local topography.

In broken country you can get a much better idea of general air drift by watching low clouds, fog, or smoke (or in clear weather, by sending up smoke puffs or small balloons) than you can by measuring a surface wind that is aerologically meaningless. Every experienced aerologist knows that in drawing weather maps and analyzing weather situations he must allow for many apparent discrepancies in surface winds.

Nor can he trust implicitly any single report of surface-wind direction or speed. The general, regional drift of lower air above the zone of surface friction is called the 'geostrophic' ('earth-circling') wind. It is also called the 'gradient' wind, because it usually blows perpendicular to the horizontal pressure gradient and because of its force-vector relations to the gradient.

For practical purposes, this gradient wind is found at an altitude of about one to two thousand feet above ordinary sea or land surfaces. It can be most easily observed and measured by means of captive balloons. If lower clouds are present at these levels, of course, they reveal the gradient wind to any observer.

TEMPERATURE AND WIND

The whole mechanism of winds alow and aloft is not a simple thing, and there is some danger of over-simplifying it in a book of this kind. But on the other hand we are certainly not interested here in the more exact and mathematical considerations of underlying causes — in such recondite things, for example, as 'entropy solenoids' between oppositely tilted surfaces of equal pressure and equal density.

In the very simplest and broadest terms, winds everywhere are fundamentally caused by differences in air temperature between two localities. The difference in temperature always causes a difference in atmospheric pressure, which in turn causes the wind. When the resulting winds are confined to small areas, not more than a few miles in extent, they blow directly from high pressure to low pressure, as one would expect.

The simplest sort of temperature-caused wind can be reproduced at will by anyone, anywhere in temperature latitudes in wintertime, and no farther afield than the front door. If the temperature in your house is around 7° F. and the temperature outside is 20° or more lower, and if you open the front door and stand in the doorway, you will notice a draft around your head. A little investigation with a smoking cigarette will reveal a pronounced draft of warm air outward through the upper part of the doorway, and a cold draft inward through the lower part of the doorway.

The draft is evidently caused by the steep temperature difference between inside and outside air, and will continue, if the door is unwisely left open, until the house is cooled to outside temperature. The thing worth noting about these miniature breezes is that, between a warm-air area and a cold-air area, the wind is from warm to cold aloft and from cold to warm below.

Expand the front-door boundary between cold and warm a few hundred times into an equally pleasant but more spacious scene — an ocean beach in midsummer — and you have the sea breeze. To study this transparently heat-motivated breeze in pristine simplicity, it would be well to pick a long, straight line of beach that marks the division between mainland and mainocean, located preferably in latitudes around 36° north where the grand planetary and cyclonic winds are feeble — say in Florida or Southern California.

Spending a twenty-four-hour day here you might notice a surface breeze-sequence somewhat as follows: early morning, calm; during the morning a breeze from seaward (the sea breeze) springs up and increases gradually until by midafternoon it is blowing ten or fifteen miles an hour; later in the afternoon it dies down, reaching flat calm at or after sunset, which holds through the evening hours; around or after midnight a fainter breeze (the land breeze) begins to blow directly off the land, increases to perhaps five or seven miles an hour during the small hours, and finally dies down to calm again well after sunrise, ready to begin the cycle anew.

Airplane by flying at three hundred feet altitude directly up or down the beach line, or parallel to it, and noting the 'crabbing' effect of the side wind, you will find that the air drift begins feebly a few miles out at sea, reaches greatest force right at the shoreline, and dies out as it becomes warmer perhaps ten or twenty miles inland.

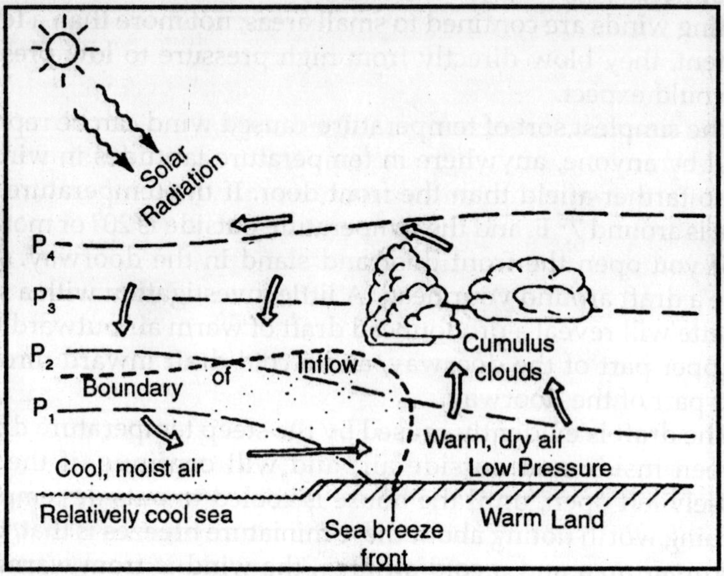

Fig. The Sea Breeze

If you now climb to three thousand feet altitude and repeat the exploration, you will probably notice an offshore counter-drift at this level, balancing the onshore surface and lower-level breeze and completing a sort of closed circulation. An airplane exploration of the night-time land breeze (somewhat difficult in practice) would show a similar closed circulation, oppositely directed at every point, and generally fainter, the surface land breeze extending only five miles or so to sea.

The reason for all these observed breezes is not far to seek. Land warms up, under a summer sun, much faster than water, and warms the air directly above it. By afternoon this warming is extended by rising currents to all the lower levels of air. As a result, the air over the land is everywhere expanded, while the air over the ocean remains much the same as it was. In the land-warmed and expanded air, surfaces of equal pressure are forced upward.

The slope of these surfaces causes, at the threethousand-foot level, say, a horizontal pressure gradient from land (upper high pressure) to sea (upper low pressure) which drives the upper counter-drift seaward. This upper counterdrift tends to pile air up over the ocean, increasing the ocean surface pressure and producing a surface pressure gradient from sea to land. This surface gradient drives the lower drift from sea to land, which earth- or water-clinging humans notice as the sea breeze.

Along temperate-zone beaches, such as those of the North Pacific or Atlantic coasts, the true sea breeze may be disguised or submerged by larger prevailing winds. Along the Jersey coast, for example, the plain daytime sea breeze, left to itself, might blow gently from east-southeast directly onshore. The general summer prevailing wind-here, however, usually blows moderately from west-southwest.

Therefore, the observed combination, or resultant wind, usually blows rather freshly from the south, or south-southeast, coming ashore only at a small angle with the beach. Along the Oregon coast, on the other hand, the daytime sea breeze may appear only as a freshening, and the night-time land breeze may show as a slight moderating, in a rather steady wind always out of the west.

Winds similar in principle to land and sea breezes, though much feebler in practice, are lake breezes and forest breezes. A large lake of course acts like a miniature ocean, with a tendency towards shoreward surface breezes in the daytime. And large forest areas, somewhat like large bodies of water, respond less quickly than more open land to daytime heating and nighttime cooling (as any aviator knows, by the air 'bumps' he strikes passing over alternating woods and open fields). As a result, a faint surface breeze blows outward from the forest in the daytime, inward towards it at night.

Still another type of heat-motivated air drift is the valley breeze. Under the summer sun the slopes and bottom of a deep valley are greatly warmed, and the air in the valley expands, pushing gradually up the mountain-sides in a drift that is usually gentle but varies greatly with mountain height and topography. At night this wind usually reverses and blows with some force down the mountain-sides, for if the sky is clear, the slopes quickly radiate away their heat and chill the air adjacent to them.

This chilled air, being denser than the atmosphere around it, pours down the hillsides and drains into all the valleys and low areas. Along the leeward (eastern) slopes of very high and steep temperate-zone mountain ranges, such as the Rockies or the California Sierra, this down-mountain night wind may whistle at near-gale force by mid-evening. And if a straight, steep canyon is fed with chilled air from a large, sloping elevated plateau, the night wind blowing out of it may be one great, roaring torrent of air. Knowing the vagaries of mountain winds, experienced campers in hilly country always build their night fire on the downhill side of camp, and sometimes use a separate daytime fireplace on the uphill side.

In all the above small-scale winds, the air flows directly along the horizontal pressure gradient from 'high' to 'low.' This observed fact would be the same for large-scale winds if this 'lukewarm bullet' of an earth on which we journey through outer space omitted to rotate as it swings yearly around the sun. But we know that the earth does rotate, making one complete sidereal revolution in four minutes less than a standard solar day. This rotation changes and complicates the whole picture of large-scale winds on earth.

PRESSURE AND WIND

At the equator, the earth's rotation amounts to the prodigious speed of more than a thousand miles an hour. In temperate latitudes — around Washington City, say — the west-east speed is something like seven hundred miles an hour. Now, suppose a large-scale outbreak, or 'mass,' of tropical air happens to drift from somewhere near the equator northward to the vicinity of Washington (a happening by no means impossible). Around Panama this air mass (unless it is blowing trees, mountains, and what-all into the ocean) is definitely moving eastward with the earth's surface at something like 1000 m.p.h.

Arriving above Washington, D.C., merely by its own ponderous inertia of millions of tons of air this errant tropical air mass should still be moving eastward at 1000 m.p.h. But at Washington the earth beneath this air mass is moving eastward at only 700 m.p.h. Is the result a wind out of the west at Washington, having a speed of 300 m.p.h., in which the National Capital is 'gone with the wind'? Fortunately, no. The only fallacy in our imaginary hurricane picture is that we have failed to take friction into account, which holds most of the winds on this earth down to reasonable speeds.

Nevertheless, the earth's rotation has a very considerable effect on the winds of this world — an effect such that any large-scale wind is inexorably deflected, to the right in the northern hemisphere, to the left in the southern hemisphere. (This 'earth's deflective force,' incidentally, is not a real force. The wind's apparent deviation results simply from the turning of the earth's surface beneath it. Nevertheless, to us humans who live on and turn with the earth, it has the effect of a real force.)

Actually, many of the winds on this earth, and most of the winds in our north-temperate zone, follow a generally circular path rather than a straight one. The traveling areas of high pressure and low pressure that bring our daily weather changes are themselves roughly circular, and usually several hundred miles wide. Variable-weather winds blow more or less circularly about these 'highs' and 'lows' — in the northern hemisphere, clockwise around the highs and counter-clockwise around the lows; in the southern hemisphere, counter-clockwise around highs and clockwise around lows.

The circular-gradient winds around a high or low, above the level of ground friction (above a thousand feet or so; that is, over ordinary terrain) move in response to forces which, strangely enough, act at right angles to the winds themselves. (But in the physical world it is really not strange for a body to move at right angles to an applied force. The moon swings ponderously along its orbit at right angles to the gravitational pull of its parent earth. A girl dancer or skater, swung in circles by her hefty male partner, acts in similar fashion.)

At the edge of a stationary or slow-moving high there are three forces acting on the winds around the high: (1) the pressure gradient pushing radially outward, (2) centrifugal force, also acting outward, and (3) the earth's deflective force acting radially inward, and exactly balancing the sum of (1) and (2). As a result, the wind has to compromise by blowing around the circle.

At the edge of a circular low, the same three forces are acting, though one of them has changed direction:

- The pressure gradient acts radially inward,
- Centrifugal force acts outward,
- Earth's deflective force also acts outward.

Thus (1) exactly balances (2) plus (3), and the wind compromises by blowing around the low. Near the equator, the earth's deflective force is relatively feeble; but here the cyclonic winds are relatively violent and their whirl is relatively small, so that centrifugal force just about balances pressure gradient. But in uppertemperate latitudes the earth's deflective force is stronger (and strongest of all, of course, at the poles), and cyclones are larger with less violent winds, so that centrifugal force is almost negligible. Hence in a large low over the northern United States, say, earth's deflective force just about balances pressure gradient.

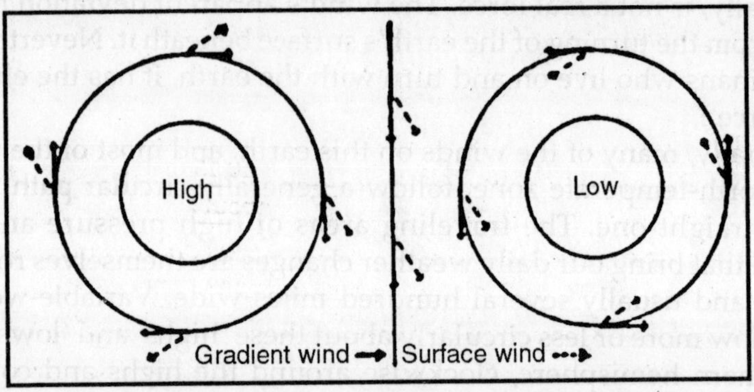

Fig. Winds Around Highs and Lows

The above statements, mind, apply most literally to nearly circular lows and to slow-moving highs. In our north-temperate latitudes most lows are not in fact circular, but are broken up by 'fronts' or distinct wind-shift lines into pie-like sectors having differently blowing and rather straight winds.

And in our latitudes, in winter, many highs drive rapidly southeastward from the Alaskan-Canadian reservoir of cold air and high pressure; one of these rapidly moving highs represents merely the motivating centre of a cold-air mass, or outbreak of polar air, and the high, gusty winds within it are likely to be mostly from the northwest. Nevertheless, basic circular-wind theory applies to all lows and highs in some degree.

Let me here repeat once again the fundamental fact of wind and pressure: *in our northern hemisphere, winds generally blow clockwise around a high, counter-clockwise around a low*. If you face directly into the surface wind, the low centre is probably to your right (or a little behind your right), and the high centre should be to your left (or a little in front of your left).

The gradient-wind speed depends on the intensity of the pressure field (which is indicated by the spacing between adjacent isobars, or lines of equal pressure as drawn on a weather map). The closer together the tenth-inch (or 3-millibar) isobars are, the stronger the gradient wind will be. As a rough but adequate guide for guessing at the gradient wind in middle-temperate latitudes, if you have even a small home-made regional surface weather map the accompanying table is useful.

So far we have been considering free air flow just above the zone of surface friction — the gradient wind. And we have seen that this gradient wind actually occurs, over ordinary terrain, at altitudes around five hundred to fifteen hundred feet.

Table. Ressure Gradient Versus Free Wind Speed

Distance in Land Miles Between Tenth-Inch (3-millibar) Isobars	Corresponding Wind Speed in Free Air
About 400 mi	10-15 m.p.h.
About 300 mi	About 20 m.p.h.
About 200 mi	About 30 m.p.h.
About 130 mi	About 45 m.p.h.
About 90 mi	About 60 m.p.h.

At that height, over ordinary land or water surface the free and unimpeded gradient wind is actually blowing very nearly at right angles to the pressure gradient and the other balanced forces. Or to say the same thing in different words, the gradient wind blows directly along the isobars. But down below near the ground or sea surface, where surface friction enters into the picture, this ideal scheme is somewhat spoiled.

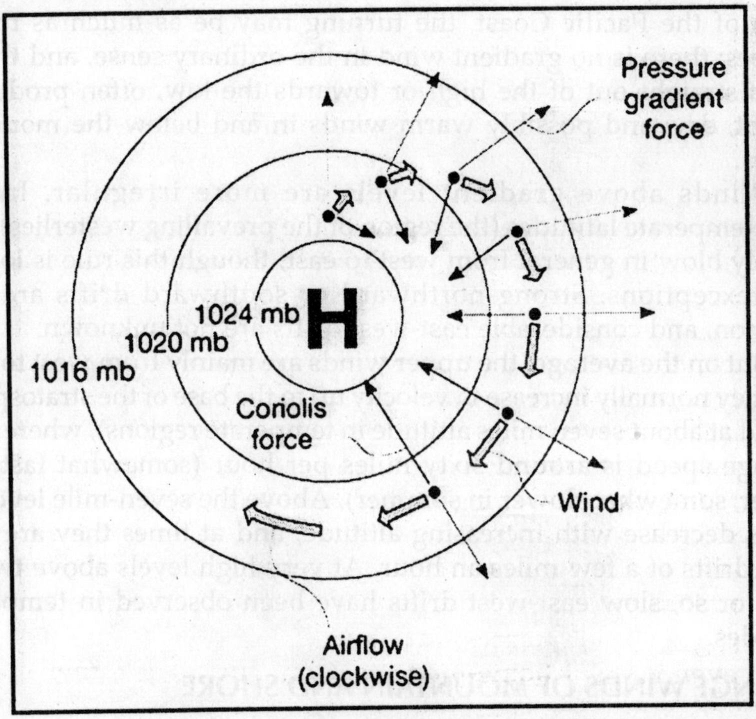

Fig. Friction Veers Surface Winds

At gradient levels (or say, roughly, anywhere between five hundred and two thousand feet altitude) the winds blow nearly parallel to the isobars. But near the earth's surface they veer *towards* the pressure gradient, because for the first few hundred feet upward, surface friction disturbs the perfect equilibrium of forces that drive the gradient winds. (Also owing to friction, the surface wind has less speed than the gradient wind.) Over ordinary terrain, this turning of the surface wind towards the pressure gradient (the gradient from high pressure to low pressure, that is) averages about thirty degrees.

As the turning of surface winds towards the pressure gradient is caused by surface friction, we should expect the amount of this turning to vary with different terrain. Such is very much the case, and the variation is extremely wide.

Over open sea or very flat country the turning is slight; the direction of the surface wind is usually within twenty degrees of the gradient, and its speed is about seventy per cent of the gradient speed. Over ordinary rolling country the turning is usually thirty degrees or more, and the speed ratio only about forty per cent.

Over very mountainous country, such as the Rocky Mountains or much of the Pacific Coast, the turning may be as much as ninety degrees; there is no gradient wind in the ordinary sense, and the air blows straight out of the high or towards the low, often producing violent, dry, and possibly warm winds in and below the mountain passes.

Winds above gradient levels are more irregular. In our northtemperate latitudes (the region of the prevailing westerlies) they usually blow in general from west to east, though this rule is loaded with exceptions. Strong northward or southward drifts are also common, and considerable east-west drifts are not unknown.

But on the average, the upper winds are mainly from west to east, and they normally increase in velocity up to the base of the stratosphere (found at about seven miles altitude in temperate regions), where their average speed is around sixty miles per hour (somewhat faster in winter; somewhat slower in summer). Above the seven-mile level the winds decrease with increasing altitude, and at times they are only mere drifts of a few miles an hour. At very high levels above twelve miles or so, slow east-west drifts have been observed in temperate latitudes.

STRANGE WINDS OF MOUNTAIN AND SHORE

It is only natural that winds are wildest and most variable on mountains. For generations mountain weather observatories have been maintained by men — those on Mount Washington, New Hampshire, and Pike's Peak, Colorado, being among the most famous in America.

And most people who love hiking, climbing, and life in the open have been caught at one time or another, during mountain expeditions, in sudden and devastating mountain thunderstorms or sudden and dangerous mountain blizzards.

When gales are rampant, any sizeable mountains are of course surrounded by terrific atmospheric eddies, whirls, downdrafts, and updrafts, which are sometimes made partly visible by drifting smoke or clouds. But visible or invisible, they are always *there,* as any aviator can testify. These mountain eddies are particularly dangerous to aviators slowing down into a normal glide for landing. I remember landing in a small but powerful Army training plane, once, in a 'pasture airport' at thirty-five hundred feet altitude on the very crest of the Blue Ridge of Virginia.

As we glided in over the treetops, we suddenly hit the lee-slope downdraft, and the 'bottom dropped out.' Only the cat-like quickness of the pilot and the 300-horsepower engine pulled us out of it in time. Flying the same little plane along the Blue Ridge windward slope on another occasion, I found the updraft strong enough to throttle back the motor to almost idling speed, gliding along the range like a sailplane at even altitude and somewhere near power-flying speed.

In the high mountains of the west, lee-slope downdrafts may have all the terrific force of 'air waterfalls.' Years ago the pilot of an old-fashioned three-motor transport, built of plywood, tried to clip too closely over a towering range in a strong wind, and hit the lee-slope 'airfall' at full speed. The ship dropped so suddenly that all the passengers (who had not even been cautioned to fasten their safety belts) were thrown bodily up out of their seats against the cabin ceiling, and the head of one of them, according to the story, actually went through the thin plywood sheeting. Luckily the wings, at least, held, and the ship got safely back to its terminal without further incident.

In addition to downward 'airfalls,' upward 'air fountains' sometimes occur. These are usually due to powerful heat-convection currents such as those which cause thunderstorms. Here is the startling experience of a mail pilot, C. E. Peebles, during a flight from Dallas to Kansas City some years ago:

In an air fountain. The wind began to boil and swirl. Suddenly and without any warning that anything so unusual was going to occur my ship started upward, completely out of control. With the nose pushed down and the engine wide open, the plane continued to go up. I had been flying at about 1500 feet. Within 60 seconds, according to the clock on the instrument board, my plane was hurled 8000 feet higher.

As the ship went up it spun much as an aeroplane does when going down in a flat spin. That was one of the ideas that flashed through my mind as it was happening. A flat spin going up! The plane was bouncing around, nose up, then down; straining every wire; being buffeted by the giant winds; completely out of control; going up, up, up. At about 10,000 feet altitude I passed out of it, or maybe it passed away from me. Anyway, the air became reasonably calm and I found that control over the plane had returned.

Among the gentler mountain breezes are waterfall winds and glacier winds. The downward air flow near a waterfall is simply air being dragged down by its surface friction with the falling water. The downwinds observed along glacier snouts are due to chilling and cooling of air over the glacier; this cooler and hence denser air drains downhill.

Much more violent are avalanche winds. Somewhat like the waterfall breeze, an avalanche wind is started by a downrushing mass of heavier material. The air is partly carried along by surface friction, but mostly pushed ahead of the descending avalanche. Cases have been observed where buildings several hundred yards beyond the zone of avalanche damage were nevertheless blown down and destroyed by the avalanche wind.

Another interesting mountain air flow is the volcano wind -breezes blowing in from all sides towards the violently ascending heated air over an erupting volcano. Somewhat the same inflowing surface breezes attend a violent detached thunderstorm, which has similar intense updrafts within its turreted clouds -and in fact, such a thunderstorm often forms over an erupting volcano, or over a localized forest or brush fire, being started by the warm, rising currents.

Some of the most famous mountain winds are great, wide, downflowing 'rivers of air' — drainage-type winds that pour down from high, relatively cold plateaus onto relatively warm adjacent parts of the ocean. There is the 'Bora' of the Adriatic Sea, pouring southwesterly down from the cold mountains of Jugo-Slavia. There is the notorious 'mistral' — a dry, cold northerly wind that pours down the Rhone River Valley in France, and out over the Mediterranean. There are cold, dry offshore gales along the mountainous coasts of Norway, and even stronger winds off the ice-covered high plateau of Greenland.

A somewhat similar wind occurs along the southern coast of Central America in the infamous Gulf of Tehuantepec. 'Tehuantepecers,' little known to laymen but veritable bogies to intercoastal mariners, occur in winter and early spring when a great southward outbreak of polar air from Alaska or Canada sweeps southward across the United States, and across the Gulf of Mexico (as a howling 'norther'), only to be partly dammed by the long mountain chain of the Central American Cordillera.

But on the Isthmus of Tehuantepec is a wide low-level pass, through which the wind blows by a sort of funnel effect which, combined with a sort of remote airfall over the mountains, scours southward out over the Gulf of Tehuantepec at perhaps sixty miles an hour. These dangerous storms give little if any local warning, barometric or otherwise, of their approach. At one moment, the weather may be mild and pleasant, with only light winds or calms. Within an hour, a northerly gale may be blowing, kicking up a steep and high sea. Even a sturdy five-thousand-ton freight steamer may lose a valuable deckload of lumber as the rising seas break over her bulwarks. Smaller vessels may be in actual danger of foundering.

The best Tehuantepecer warning for weather-wise mariners, incidentally, is radio ship-weather reports from the Gulf of Mexico, north of the Isthmus — if there is no norther up there, no Tehuantepecer is possible. One of the strangest and most spectacular mountain winds, though nevertheless widely useful, and often even pleasant to winter-chilled beings, is the foehn wind or chinook. Along the leeward (eastern) slopes of the Rockies, this wind occasionally develops in such spectacular form that it is called the 'snow eater.' When this great foehn blows, the temperature may rise by as much as *40° F. in a quarter of an hour*, the humidity tumbles sharply, and snow disappears as if by magic.

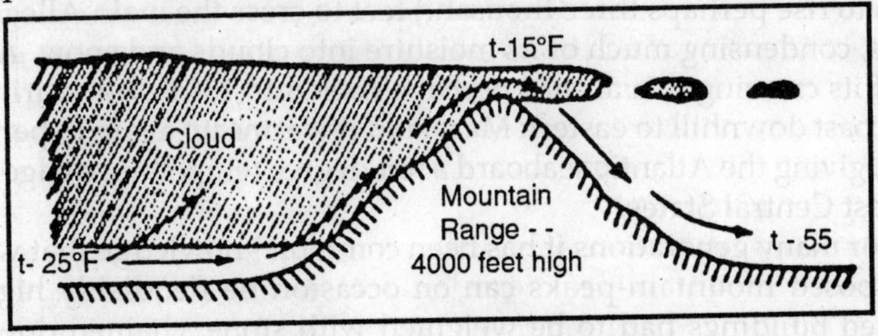

Fig. Mechanics of the Foehn Wind

The interesting and beneficent mechanism of the foehn wind is somewhat as follows: When a strong general air flow strikes the windward slopes of a mountain range, the air is forced to rise over the range. If humidity is high, clouds are soon formed by a mechanism. Of course the rising air expands, hence cools. But after cloud-forming begins, the rate of cooling with altitude is only about 2 F. for each rise of 1000 feet. On this upward climb the air loses most of its moisture in the form of rain or snow.

Topping the range, the air promptly starts its long glide down the leeward slopes. Here it is warmed by compression as it sinks to lower levels — but warmed, since no moisture transformations are here involved, at the rate of about 5° F. per 1000 feet altitude. Finally arriving again at its former flat-country level, the air (having warmed twice as fast as it cooled) is naturally at a higher temperature than before. How much higher depends on the height and general topography of the mountain range, and on the general speed of air flow. Also, as the air may follow paths well above the surface, the warming may be more than the mere topographic height of the mountain range would indicate.

Until about 1933 the dry, searing northerly 'Santa Anas' of Southern California were thought to be warm because of their desert origin. But scientists at California Tech. proved that during these winds, temperatures on the mile-high Mojave Desert were actually much lower than those on the coastal plain around Los Angeles. Hence the Santa Ana and other warm, dry northerlies in Southern California are in fact true foehn winds whose warmth is due entirely to descent-compression heating.

Though not generally recognized as such, a powerful foehn effect moderates the midwinter climate of the Eastern seaboard during vigorous polar-air outbreaks, or 'cold waves' that bear down out of the northwest. In crossing the Alleghenies between Pittsburgh and Washington, for example, the same air that sets Ohio shivering at 25° F. has to rise perhaps three thousand feet to cross the main Allegheny ridges, condensing much of its moisture into clouds and snow. At the top of its crossing this air has perhaps cooled to 15° F., but during its long coast downhill to eastern Maryland it warms up again to perhaps 35° F., giving the Atlantic seaboard a ten-degree temperature edge over the East Central States.

For many generations it has been common knowledge that winds on exposed mountain-peaks can on occasion be fiendishly high — exposed buildings had to be weighted with stone, chained down at the corners, and so on, or they would simply blow away. But only in recent years have these icy gales on lonely peaks been measured with heated anemometers designed to resist the choking action of common mountain snow and rime and icing. On the twelfth of April, 1934, a modern mountain anemometer happened to be exposed above the Harvard Observatory on Mount Washington in northern New Hampshire.

Now, Mount Washington is only a 6000-foot peak, such as would scarcely be dignified with the name of 'mountain' in the more perpendicular parts of the Great West.

But the convergence of cyclone tracks through New England puts Mount Washington squarely in the groove followed by many departing storms, and makes it climatically equal to much higher Pike's Peak. On this pleasant April day in New England high southwesterly gales were general as a northeastward-hurrying tropical air mass felt the advance crowding of an oncoming cold front and politely tried to get out of the way.

Even at surface stations the southwesterly wind was blowing fresh to strong. During most of that afternoon the wind on Mount Washington summit was well above a hundred and fifty miles per hour; and at one time it reached the *officially verified all-time record velocity of two hundred and thirty-one miles an hour.*

Small Balloons and Large Breezes

Clouds have been for ages man's chief guide to the direction and force of the upper winds — by day under the sun, by night under a high moon or bright stars. Even on a cloudless night, you can get some idea of wind force in the upper levels from the twinkling of the stars — the harder it blows, the more they twinkle.

Lower clouds from five hundred to two thousand feet above the ground are particularly useful because they serve to show gradient winds.. In the absence of lower clouds, gradient winds can be found most quickly and easily by means of a small smooth-bore mortar similar in principle to those used for exhibition fireworks. The mortar fires a small, cheap pasteboard shell, loaded with powder, straight upward. By means of a time fuse the shell bursts at five hundred to a thousand feet altitude into a compact puff of white smoke that drifts with the wind long enough to be noted by an observer directly beneath it.

You can observe the drift-direction of clouds (which are, of course, imprisoned within the upper winds that bear them) by looking straight upward at them along a vertical pole having north-south and east-west cross-arms at its top. And by considering a cloud's altitude and carefully watching its drift, you can form some estimate of its speed. Suppose you watch, say, a cloud passing directly overhead, and time its progress between

- The zenith, or a point directly above you,
- A point some twenty-six and one half degrees (measured roughly by fingerspread) below the zenith.

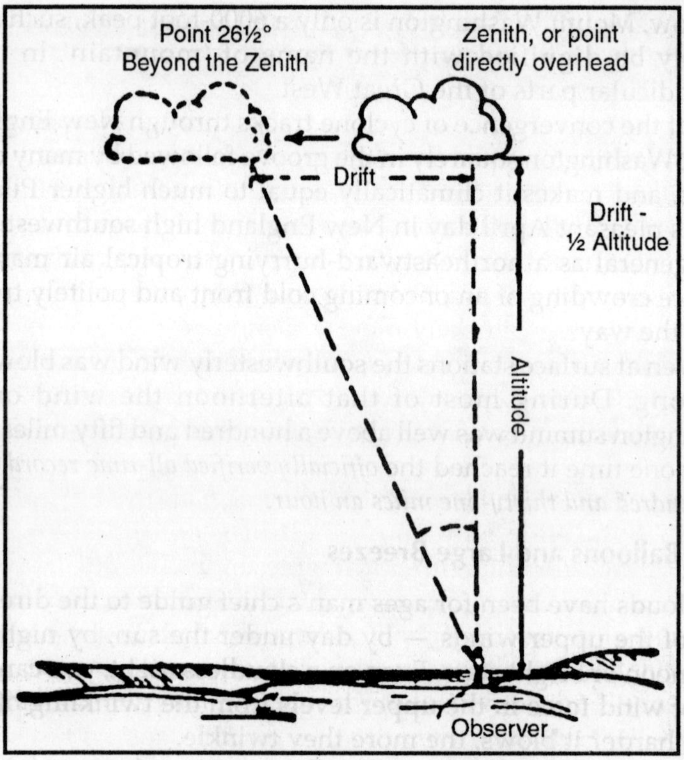

Fig. Gauging the Speed of Cloud Drift

During this time the cloud has traveled a horizontal distance equal to half its altitude. It is not uncommon to note wispy cirrus clouds or 'mares' tails,' for example, which usually float at six or seven miles altitude, hurrying along at speeds better than a hundred miles an hour. In mountainous country that is familiar from exploration or careful map-reading, you can often judge the direction and speed of detached clouds by watching their shadows.

There are also more refined and accurate ways of measuring the drift of clouds. Most widely used has been the nephoscope — a horizontal circular mirror of dark glass which reflects the sky and the clouds. This reflection is viewed through an eyepiece mounted above the mirror in such a way that it can be moved around its rim. Suitable scales are etched on the mirror itself, and the eyepiece is adjustable for height. Hence cloud height and speed are represented in miniature by the principle of similar triangles.

Another device for gauging cloud drift overhead is a vertically aimed right-angle telescope having a large field of view and low magnification, as well as suitable cross-hair scales in the eyepiece itself and a circular azimuth scale around which the whole instrument may be rotated. Still another instrument is a cameraobscura arrangement (which may be nothing more than an ordinary reflex camera pointed straight upward), also rotatable around all points of the compass and provided with suitable scales on its ground-glass focusing plate.

The most accurate device for checking the direction and speed of a cloud (or any other object in the free air) is the theodolite.

The instrument itself — a free-air development of the surveying transit — is simply a small right-angle telescope, magnifying perhaps twenty times and accurately mounted on azimute-angle (horizontal) and elevation-angle (vertical) circles. Wherever the objective tube of the theodolite may point in following a drifting cloud or balloon, the eyepiece tube is always horizontal and at a convenient height for the observer's eye. Azimuth-angle and elevation-angle motion of the objective tube are controlled by slow-motion knobs and worm gears, and both the azimuth and elevation scales can be read by means of verniers to the nearest tenth of a degree.

For accurate measurements of cloud drift two theodolites are required, located at the ends of a surveyed base line perhaps a mile long. If the two observers can agree by telephone or radio telephone on what cloud (or what part of a larger cloud) they are following, they can obtain by ordinary triangulation a double-checked determination of the cloud's height, direction, and speed.

But if you are going to all the trouble of theodolite observations, it is usually simpler and more certain to employ as wind-riders, not the large and usually loose clouds of Nature, but those much smaller man-made clouds of hydrogen or helium prisoned in a bubble of thin rubber that are called pilot balloons.

A pilot balloon leaves the ground as a glistening, bubble-like sphere some two feet across, and swells gradually as it rises. At last, grown after an hour's time to six-foot breadth in the thin, cold air of the lower stratosphere, it ignominiously but heroically bursts. Its drift at all levels, however, has been followed by one or more observers with theodolites.

The balloon itself resembles a child's toy balloon in general construction and appearance, though more than twice as big. Perhaps six inches in diameter uninflated, it is gradually blown up with hydrogen or helium to a diameter of 26 inches or so before being released. The hydrogen or helium comes from a heavy, boy-size pressure tank that holds enough gas for twenty or more balloons. Hydrogen entails a slight fire hazard, but it is much cheaper, and being lighter, is more efficient.

In practice the balloon is inflated (with hydrogen) until it will just lift off the ground a weight of four and two-thirds ounces — until its 'free lift,' in other words, is 4.66 oz. (If helium is used, the free lift is increased by ten per cent to compensate for the slightly greater density of this gas.) Under average atmospheric conditions a sound standard balloon inflated to this free lift will rise at a fairly uniform rate of about 200 yards per minute. This rise rate is standard for all ordinary pilot balloon work and it is an important element in the plotting technique.

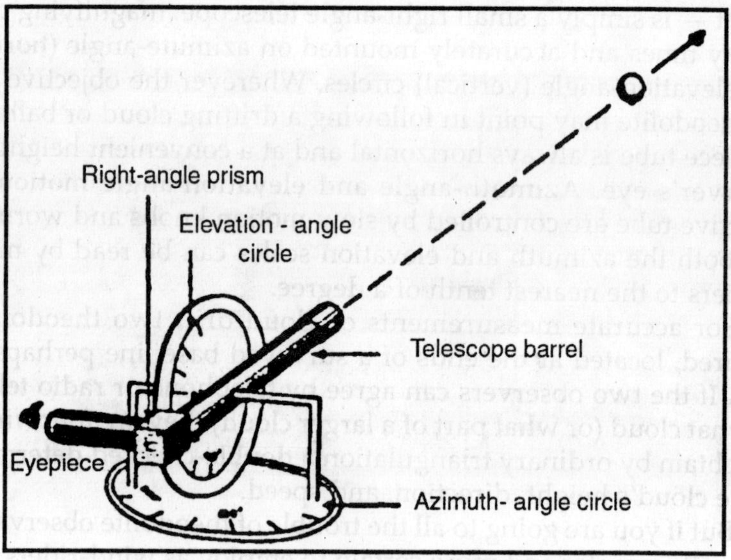

Fig. Sighting a Theodolite on a Pilot Balloon

For use under clear skies, pilot balloons are either of uncolored latex (which gleams yellowish white and almost pearl-like in bright sunlight) or, better, tinted with aluminum pigment that gleams even brighter with a silvery tinge. For an ascent under high, thin, well-lighted clouds, the balloon is usually colored yellow or red. But when the clouds are dark, opaque, or lowering, a black-colored balloon is easiest to follow with the theodolite, and black is also the best colour just after sunset or before sunrise.

At night the standard pilot carries up with it, at the end of a few feet of string, either of two different light sources — candle lantern or electric light. The first is a small, cylindrical paper lantern about three and one half inches wide by seven inches long, in the bottom end of which is fixed an inch or so of ordinary three-quarters-inch candle that will burn for half an hour.

The whole lantern weighs only half an ounce, and the balloon is, of course, overinflated by this amount in order to maintain the standard rise rate. The other light source, used in forest and other regions where fire hazard is something to be feared, is a small 2.3-volt flashlight bulb connected to two flashlight cells of the smallest type, the whole weighing about one ounce.

Though the electric light is perhaps somewhat safer, and certainly easier to launch in a high surface wind, than the candle lantern, it gives out the latter's one candlepower only when first lighted, and thereafter continuously grows dimmer with the constant exhaustion of its battery until, after half an hour, its feeble glow is a mere quarter-candlepower. Also, the electric light is considerably heavier, and makes such a solid and compact missile that it might conceivably hit an unprotected human head with lethal force on dropping from a height of six miles or so after the balloon bursts. Hence it should usually be provided with a handkerchief-sized safety parachute.

Though ordinary pilot balloons rise for an hour or so before bursting, and both electric lights and candle lanterns burn for more than half an hour, the balloon's drift is seldom followed with ordinary theodolites for more than half an hour in the daytime or twenty minutes at night. The reason is that, as the balloon rises higher and higher, it usually encounters stronger and stronger winds and blows rapidly away to greater and greater distances.

Towards the end of a night run under clear skies, either the candle lantern or the electric light resembles a very faint star which can be discerned only with increasing difficulty as it appears to move among the other stars, and is finally lost even by a desperately straining eye. In round numbers, the little light can be followed to 'distances out' of three to six miles, corresponding to altitudes (in average upper winds) of six thousand to possibly fifteen thousand feet above the observing station.

Under clear daytime skies the far-distant and receding balloon, seen through the theodolite, looks like a minute white speck drifting slowly (though it may be moving at more than a hundred miles an hour) in the upper blue. As the elevation angle gets lower with increasing distance out, also, the sky gets whiter and more hazy, making the elusive white speck still harder to see. A red filter on the theodolite (similar to the red camera filters used in most air, sea, and mountain photography) helps somewhat by making the balloon appear as a bright red speck against a dark field. But even so, the balloon is usually lost, even by expert observers, at a distance out of ten or twelve miles, corresponding (in average upper winds) to an altitude of fifteen thousand to possibly thirty thousand feet.

Of course there are larger pilot balloons that can be seen by day to greater distances out and higher altitudes. Formerly these larger balloons were made by cumbersome processes. As a result, they were prohibitively expensive and heavy. But around 1930, the Dewey and Almy Chemical Company in Cambridge, Massachusetts, developed a process of manufacture in which the rubber-latex balloon, while still soft, is air-expanded to its permanent (uninflated) size. Thus came about better balloons that cost very little, and are light enough to rise rapidly. There are now available, also, more powerful balloon light sources that can be seen farther at night.

Of course any pilot balloon, if it rises into or above clouds, is gone with the wind' for all the good it does an ordinary ground observer. For this reason, high upper-wind observations are at present confined almost entirely to clear weather, and a common report along the airways is, 'No upper winds due to low clouds,' or 'Balloon entered clouds at 5000 feet.' But by using directional radio — a miniature transmitter that goes aloft with the balloon and directional receivers on the ground — it is possible to overcome even these difficulties.

PLOTTING THE UPPER WINDS

Balloon-gazing, though perhaps a childish and lazy pastime, is a pleasant one for most people and never lacking in a certain fascination. But if we are to fathom the complex and viewless veerings of the upper winds, mere balloon-gazing is not enough.

Though plotting balloon runs is simplicity itself in principle, trained men are needed for quick and accurate performance. There must be one qualified *observer* (plus an extra man to read the angles on long, high runs) at each theodolite. There must also be a qualified *plotter* at the plotting board, who is at all times connected with the one or more observers by telephone. Each man wears a telephone set of the operator's type, so as to leave his hands free.

In front of the plotter is a stop-watch or clock with a large sweep second hand. As the observer drops the balloon he calls 'Gone!' (through the telephone), and simultaneously the plotter starts the timer. Thereafter, at five seconds before each exact minute, the plotter calls 'Warning!' and the observer takes care to centre the theodolite cross-hairs exactly on the drifting balloon. On the exact minute the plotter calls 'Read!' and the observer momentarily stops turning the slow-motion screws on the theodolite, and reads the azimuth and elevation angles as they are.

Noting the readings of the angular scales and verniers, the observer calls, for example, 'Azimuth one nine four point four [194.4° clockwise from north, or about SSW], elevation five four point one [54.1° above the horizon].' In order to eliminate misunderstandings, the plotter repeats back these angles as he records them and plots on the plotting board what amounts to a map of the course followed by the balloon. The plotting board is a large table-top perhaps three feet or so square. At the centre is a pin, from which, as a central pivot, a long metal arm can be swung to any desired reading on a circular azimuth scale around the outer rim of the board.

Having set this arm to the announced azimuth for any given minute, the plotter refers to a table giving the balloon's (presumed) altitude for that minute. Knowing this altitude and the angular elevation, he quickly computes the balloon's horizontal 'distance out' by slide rule (a simple tangent relationship), or gets it from a table. This distance out, in yards, he lays off on the scale engraved on the metal azimuth arm, and is then able to plot the horizontal projection of the balloon's exact position in space at that particular minute as a pencil point on the board.

As the minutes go by, the whole series of these pencil points on the plotting board will show a clear 'map view' of the horizontal drift of the balloon in response to the varied upper winds encountered by it at increasingly high levels. Having completed the run, the observer goes back over the series of plotted points in order to find the wind direction and speed at all the altitudes through which the balloon ascended.

In order to get the upper wind at 3000 feet, say, the plotter refers first to the pencil point representing the balloon's position at the fifth minute. (At a rise rate of 200 yards per minute, the balloon reaches 3000 feet altitude at end of the fifth minute.) But in order to get the average wind at 3000 feet, the plotter actually notes the plotted balloon drift between pencil points corresponding to the fourth and sixth minutes. With a suitable celluloid protractor, which may be graduated in compass points such as 'north-northeast,' or in azimuth degrees geographic (measured from true north), the observer notes the wind direction between the fourth minute and sixth minute. This noted wind direction is, in effect, the actual wind direction existing at the fifth minute, or at 3000-feet altitude, during the time when the balloon run was made.

In order to find the wind velocity at, say, 3000 feet, the plotter simply measures off the plotted distance between fourth minute and sixth minute with a suitable celluloid scale (usually combined with the protractor) which reads directly in miles per hour. The wind direction and wind speed at various other desired levels are found in exactly similar fashion.

Thus by the simple operations described above a trained plotter can construct a comprehensive picture of all the varying winds above his head just as quickly as the balloon itself rises through these winds — up to a height of say fifteen thousand feet in less than half an hour. Single — theodolite plotting technique is so simple and efficient that it is in standard use today all over the United States and in many other countries. Yet the whole accuracy of the method depends on a sort of Achilles heel — the assumption that the balloon will maintain its allotted rise rate.

In perhaps nine cases out of ten, the balloon does dutifully rise at just about two hundred yards per minute. But under certain aerial conditions, such as heat-caused updrafts in mid-afternoon or sudden downdrafts near a cold front, the balloon may (all unknown to the observer) decline to oblige. In other instances, fortunately rare, the balloon may spring a slight leak after leaving the ground. Besides decreasing the balloon's rise rate and spoiling the accuracy of a single-theodolite upper-wind observation, such a leak may incidentally give the balloon itself a new lease on life and a long journey in the high-level hurricanes of the substratosphere.

One such balloon was released from Havre, Montana, on a cloudy October day in 1934. Becoming an aimless floater in the substratosphere, it traveled southeastward the remarkable distance of nine hundred and thirty miles before being picked up, two days later, on a farm in Iowa.

The upshot of all this is that no aerologist trusts any singletheodolite balloon run too implicitly unless the balloon's actual height at certain times has been checked by independent means. The most thorough and accurate check on the balloon's actual altitude from minute to minute is obtained by a double-theodolite observation similar to that used for the accurate tracking of clouds. The two theodolites are set up at the ends of a surveyed base line at least a mile long and so far as possible perpendicular to the line of expected balloon drift.

The two observers are connected by telephone or radio telephone with each other and with the plotting board. By triangulation similar to that used in ordinary topographic surveying, the balloon's true altitude and distance out at the end of each minute can be accurately, if rather laboriously, computed. Where double-theodolite or even triple-theodolite balloon runs are the rule, as at certain large military balloon research stations, it is possible to design a rather complex plotting board that will quickly and graphically solve all the necessary triangulation.

Even when using only one theodolite, an observer can get

independent checks on the balloon's distance away from him (and hence, knowing the elevation angle, on its actual altitude). He can sight on it with a standard yard-wide military range-finder similar in principle to the miniature distance-finders used on precision cameras. But holding in the field of a range-finder the image of that rising, bouncing, and drifting balloon is no easy task.

A better scheme includes a white-paper target (say a very light open one-by-three-foot cylinder of tissue paper on a wire frame) hung by a piece of string at a convenient known distance (say twenty or fifty yards) below the balloon. Balloon and target are sighted on with a military binocular equipped with a 'mil-scale' — or the mil-scale, or an adjustable angle-measuring cross-hair, may be included in the theodolite itself.

The mil is a useful unit of angular measure — small (about eighteen mils equal one degree), and convenient (one mil at the target equals one thousandth of the range). Thus by a very simple computation (allowing for the foreshortening of the string due to the balloon's elevation angle) the balloon's distance away is obtained.

The greatest peace-time user of upper-wind data, not only in the United States but all over the world, is aviation. Every transport air-liner captain, every military or naval officer flying a sizeable warplane, and in fact every pilot, professional or private, who takes his cross-country flying at all seriously — every last aviator, in short, who wants to have some idea of favorable flying levels for a given journey as well as what time he is likely to reach his destination — makes a thorough study of all the available upper-wind reports along his route before taking off.

Moreover, he listens regularly to such radio broadcasts of upperwind reports and other weather reports as he can receive during the progress of the flight, and he may even call a ground terminal.

If permitted by Government airway regulations, an aviator may alter his altitude from time to time in search of that much desired, beneficent 'tail wind,' as opposed to the malevolent 'head wind' that holds him back and wastes his gasoline. Aviation upper-wind reports are usually listed in miles per hour and degrees of azimuth (from true north), so that they can be applied directly to the customary course-steered-upper-wind-coursemade-good diagram used in laying out a well-planned crosscountry flight.

Aside from aviators, aerologists themselves are the most numerous peacetime users of upper-wind reports. Every important weather-forecasting station prepares at least two upperwind maps daily, showing in one integrated picture all the upper winds, at various places and at all available levels, over a wide region around the station. These maps greatly aid the weather forecaster in noting and anticipating the movements of air masses, high-pressure areas, cyclones, and fronts.

In wartime the artillery is also a great user of upper-wind data. Even 6.2-inch corps artillery commonly fires its shells, en route to the enemy lines, up to altitudes around three miles. Heavy (14-inch) railroad-artillery shells may ascend to ten miles or more if fired at a high angle. As an extreme case, the ponderous and spectacular but wasteful and inefficient 'Big Berthas' that bombarded Paris in 1917 from seventy-odd miles away flung their high-velocity 8-inch shells up to heights around thirty miles. Anti-aircraft shells also, of course, may on occasion be fired to considerable heights — say up to five or six miles.

Upper winds act just as forcibly on a heavy cannon shell fired through them as they do on the lightest of pilot balloons — but the shell travels so fast that it does not give the wind very long to act. The effect of unknown upper winds may be enough to cause wide misses instead of direct hits, thereby wasting large shells which cost (with their propelling charges) hundreds of dollars apiece.

Therefore every important artillery unit is constantly taking its own pilot-balloon runs, and finding all the various upper winds that a shell must encounter in its arching flight. By weighted vectorial addition, the actual winds at various levels are combined into one average or 'ballistic' wind. Correction for this ballistic wind is applied to the gun sights along with corrections for air density and other factors.

Chapter 5

Visible and Invisible Moisture

Invisible water vapour is always present in the air, varying in amount from a mere trace up to about four per cent by volume or two and one half per cent by weight (the water vapour being somewhat lighter than dry air). 'Specific humidity' — the actual mass of water vapour present in a given mass of air, measured usually in grams of water vapour per kilogram of air, and varying from less than 1 g/kg in very dry air up to perhaps 20 g/kg in very moist air — is the most conservative and generally useful measure of atmospheric moistures.

For some purposes, though, dryness or moistness of air is more conveniently expressed as 'relative humidity' — the ratio of the actual amount of water vapour present (the existing specific humidity) to the greatest possible amount of water vapour that could exist at the existing temperature (the saturation specific humidity). And for many practical weather purposes, the best measure of atmospheric moisture is the 'dewpoint'-that temperature which has a saturation specific humidity equal to the present, existing specific humidity. Finally, we saw how any one of three conditions indicates that condensation of water vapour into liquid form is imminent:

- The nearness of the existing specific humidity to the saturation specific humidity;
- The nearness of the dewpoint to the temperature;
- The nearness of the relative humidity to one hundred per cent.

When water vapour in the air comes out of its invisible form into the visible form of liquid water, however finely divided, it must have some solid nucleus on which to form. Now, on a summer night, say, if the lower air is quite still as the dark earth radiates away its heat, the cooling ground cools only that layer of air immediately next to it.

When the temperature of this still, surface air reaches its dewpoint, invisible moisture condenses into liquid dew on sticks and stones, blades of grass, or any other cool and convenient objects — nuclei of condensation, we might call them, albeit large ones. Thus the fairly common idea that 'dew falls' is wrong. If the temperature happens to be below freezing (less than +32° F. or 0° C.), the invisible water vapour sublimes directly into feathery ice crystals, forming hoar-frost.

Invisible water vapour that condenses into minute droplets in the free air, forming fog or clouds, also needs solid nuclei on which to form. But whereas the nuclei for dew and frost are large and evident, nuclei for fog and cloud droplets are of infinitesimal size. Nevertheless these nuclei — minute particles of sea salt, smoke, hygroscopic dust, and the like — are always present in the troposphere.

Given these nuclei, rapid condensation of invisible moisture into minute droplets of liquid water (visible in the aggregate as fog or cloud) begins when, by reason of evaporation, dewpoint increases to existing temperature; or when (as is far more usual), by reason of atmospheric cooling, temperature reaches existing dewpoint. In any case, condensation begins when and where dewpoint and temperature come together, for any reason, at any level in the air ocean.

FOG

Fog can be at best a creation of elfin beauty, is on the average an inconvenience scarcely ever welcomed by honest people, and becomes at worst, in transportation, a great hazard. It may stall land vehicles for hours, force ships along the coast to anchor for days, and even more sorely bedevil airplanes that must keep moving at two or three miles per minute if they are to keep the air at all. In temperate latitudes the gray, creeping fog-mass is usually formed entirely of myriads of minute water droplets, each something like one thousandth of an inch in diameter.

Damp haze, which in its denser forms may approach the murkiness of light fog, can (unlike fog) begin forming when temperature and dewpoint are far apart. This denser form of haze sometimes extends for several thousand feet above the earth's surface, and the well-defined top of the haze layer usually marks the base of a temperature inversion.

To an aviator above the haze it appears almost as a level, pearly-white sea brilliantly reflecting the sunlight; only as he descends into the haze do ground patterns begin to loom murkily into view. But at night, lights shine upward clearly through this haze, whereas true fog of any considerable thickness blots them entirely from view.

The reason for this is that fog droplets are much larger than damp-haze droplets (though still incredibly small) — large enough, in relation to the wavelength of visible light, to block off almost completely any direct rays. Yet the transition from dense haze to light fog may be gradual, with gradually increasing relative humidity. As dense fog requires both plenty of vaporous moisture to start with, and then considerable cooling, for its formation, it is found mostly in temperate latitudes. In arctic regions, or at high altitudes, fog (or cloud) may be composed of myriads of minute ice-needles — 'frost smoke' that glitters and sparkles in the sun. In the tropics fog occurs but rarely, because the temperature contrasts that favour it are usually lacking.

Fog is classified, according to the mode of its formation, into three main types:

- *Radiation fog,* also called 'land fog' or 'summer fog,' in which air already stagnant on the scene cools until its dewpoint is reached;
- *Advection fog* or transportation fog, also called 'winter fog' or 'sea fog,' in which air is brought in bodily from outside regions to the scene of its fogginess;
- *Frontal fogs,* usually temporary and transient, that occur with the passage of fronts between unlike air masses.

Radiation Fog

Radiation fog is most frequent during the long, clear nights of autumn, though it may occur at any season, particularly in rather stagnant air conditions near the centre of a high-pressure area. The fog sometimes appears to rise from the ground; but actually, the only thing that rises is a zone of temperatures lower than the dewpoint of the air in question.

As the ground and the moist air immediately above the ground slowly radiate their warmth away into space, the temperature of this lower air falls, approaching its dewpoint. When the two coincide, fog appears in visible form. As the top of the fog layer itself radiates away heat much faster than dry air, the fog usually continues to deepen until sunrise.

On a flat plain, if the wind is very light (say only one or two miles per hour), the fog layer may be very shallow. This 'ground fog,' perhaps only a few feet thick, results from very gentle mixing in the air immediately above the ground; there is not enough eddying and turbulence to raise the surface coldness higher. But in rolling or mountainous country, even on a calm night, cooled surface air drains into the hollows (as anyone can notice by driving over darkened hill and dale in an automobile, or even walking across broken terrain, with a thermometer). As a result, large pools of valley air may be cooled sufficiently to produce fog 'lakes' many fathoms deep.

If the breeze is somewhat stronger — say two to six miles per hour

— the surface coldness is mixed by eddying and turbulence through a considerable layer of the lower air, causing true fog that blots out hillocks and hollows alike. This sort of fog usually begins at or near the surface, working gradually upward, though to an observer on the ground it seems to materialize rather suddenly out of nothing more tangible than a few stray wisps of vapour.

The maximum height of the fog depends on many aerological factors, but it is seldom more than a few hundred feet. Temperature in the fog layer usually decreases with altitude at something like the moist adiabatic rate (-2½°/1000 ft); and just above the top of the fog there is usually, if not always, a temperature inversion.

If the breeze is much stronger than six miles an hour or so, eddying and turbulence diffuse the surface coldness through such an extensive layer of overlying air, and bring down so much potential warmth from aloft, that the air temperature never reaches its dewpoint and fog is entirely prevented.

The conditions favorable to deep and bothersome radiation fog are, therefore: to begin with a fairly high dewpoint for the season (plenty of actual water present in vaporous form, that is), and not much difference (say less than 15° F. at sunset) between temperature and dewpoint (or a corresponding small difference, say less than 7° F., between wet- and dry-bulb temperatures); second, a clear or nearly clear sky, for maximum radiational cooling; and third, through at least part of the night, light breezes registering two to six miles per hour at the surface.

Once begun, radiation fog continues to deepen through the hours of darkness and, curiously enough, may increase for a brief time after sunrise. This last effect may be due to either of two factors: the power of sunlight to make floating particles attract water; or the increase of lower-level turbulence and mixing caused by solar heating at the surface. It is a rather frightening experience to take off from a clear airport at dawn, and see it fog in under you a few minutes later.

Within a few hours after sunrise, however, an ordinary radiation fog begins to 'burn off.' This burning off always proceeds from the ground upward as the solar radiation, penetrating through the fog layer, warms first the ground itself and then, progressively, the air above it. The lower fog does not actually 'lift' in this case any more than it 'rises' out of the ground at night; what actually lifts at first under the morning sun is a zone of temperatures higher than the dewpoint.

But before long rising currents from the warmed ground break through the fog layer here and there as the solar radiation continues. And the sun, shining through these breaks with redoubled energy, soon produces enough warm updrafts to raise the whole fog layer slowly upward and sweep it into scattered drifts of cloud.

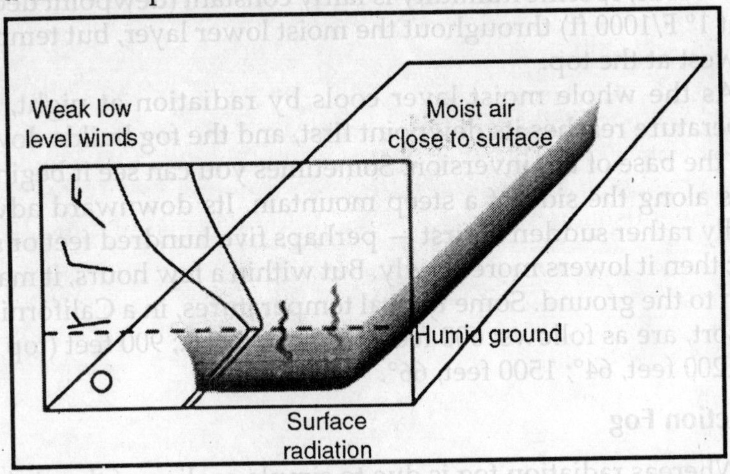

Fig. Radiation Fogs

The worst radiation fogs are found around great cities, where innumerable smoke, oil, and soot particles impair visibility on their own account, serve as ready nuclei for condensation, and hinder the evaporation of fog once formed. Thus the infamous London fogs, which greatly increased during the late nineties with the growth of the great city, may hang on in yellowish 'pea-soup' form, perhaps for days, until the advent of a new air mass and freshening winds.

Though less tenacious, radiation fogs are nearly as common in many regions of the open country. They can also occur over limited areas of calm ocean, but the usual sea fog arises out of different causes. An extreme example of radiation fog occurs in the Great Valley of California, where murky-gray obscurity may hang on for days beneath the 'lid' of a temperature inversion.

A particular type of radiation fog, called 'high fog' or 'stratus fog,' starts in mid-air and builds downward. This high fog, also, is particularly prevalent in the walled-in valleys of California, but I have noticed it occasionally over the eastern seaboard, and it probably occurs widely, though rather rarely, in other regions. Once we had a formation of army airplanes operating all night from an air base in the District of Columbia. This high fog formed suddenly, around midnight, at a height of about fifteen hundred feet, trapping one of the planes above it.

The predisposing aerological condition here is a zone of decreasing temperature and high humidity from the ground upward for the first thousand feet or so, with a pronounced and lid-like temperature inversion above. The air below the inversion is thoroughly mixed, equally moist, and coldest at the base of the inversion in mid-air. In other words, specific humidity is fairly constant (dewpoint decreasing about 1° F/1000 ft) throughout the moist lower layer, but temperature is lowest at the top.

As the whole moist layer cools by radiation at night, the top temperature reaches its dewpoint first, and the fog builds downward from the base of the inversion. Sometimes you can see it beginning in wisps along the side of a steep mountain. Its downward advance is usually rather sudden at first — perhaps five hundred feet or so in an hour; then it lowers more slowly. But within a few hours, it may reach down to the ground. Some typical temperatures, in a California fog of this sort, are as follows: 600 feet (surface), 57° F.; 900 feet (top of fog), 54°; 1200 feet, 64°; 1500 feet, 66°.

Advection Fog

Whereas radiation fog is due to simple cooling of the air in place, advection fog may be due to either of two causes. One is the horizontal advance of warm, moist air over a cold land or water surface. The other is the advance of cold air over a moist, warm surface. Of these two, the first cause is by long odds the most common.

On land, a warm-air advection fog often results when gentle, moist southerly winds blow in from the ocean, particularly if the land surface is covered with snow. The warm, moist air is chilled by the cold ground below it until its falling tempera- ture, uniformly distributed through several hundred feet of depth by gentle turbulence, reaches its dewpoint. In the United States any large northward outbreak of warm, moist (tropical maritime) air in wintertime may produce widespread advection fog.

The warm sector of a marked temperate cyclone may be completely blanketed by such a fog, and if the pressure gradients are steep, this may be one of the rather exceptional cases where fog accompanies a high wind, usually from some southerly quarter. These warm-air advection fogs are far less common on land than radiation fogs; but the advection fogs are likely to be more widespread and prolonged, forming or advancing at any hour of the day or night.

On the ocean, warm-air advection fogs are by far the most common type, and more frequent in summer than in winter. They are likely to form wherever a cold ocean current or area is surrounded by considerably warmer water — such as off Newfoundland, where the Labrador current projects southward; off Nantucket, where the colder water from the northern New England coast meets the warmer waters from south of Cape Cod; or along the whole line of the Pacific coast, where water upwelling inshore from chill depths below is much colder than the Pacific waters offshore.

The cold-water and fog area may be small and distinct — such as the five-mile-wide cold current in the middle of Santa Barbara Channel. The whole fog area moves with the prevailing winds, which are usually light -say fifteen miles per hour or less on the open water — though in a great temperate cyclone they may be much higher. I remember one case of moderately thick fog on a steamship off the Atlantic coast, with the wind blowing a whole gale out of the southeast and a mountainous sea running. If the wind is much above twenty miles per hour, however, the minute fog droplets tend to coalesce into the larger droplets of mist or drizzle, with somewhat improved visibility.

These sea fogs vary widely in height — sometimes they are so shallow that the mastheads of ships project into clear air above. At other times they are more than a thousand feet deep. Their average depth is perhaps around five hundred feet. In any actual case, of course, this height can be determined by a kite or captive-balloon sounding from a ship in the fog — or from shore elevations along a mountainous coast like California. Sometimes the sea murk may take the form of 'high fog' or low stratus, with good visibility along the water surface.

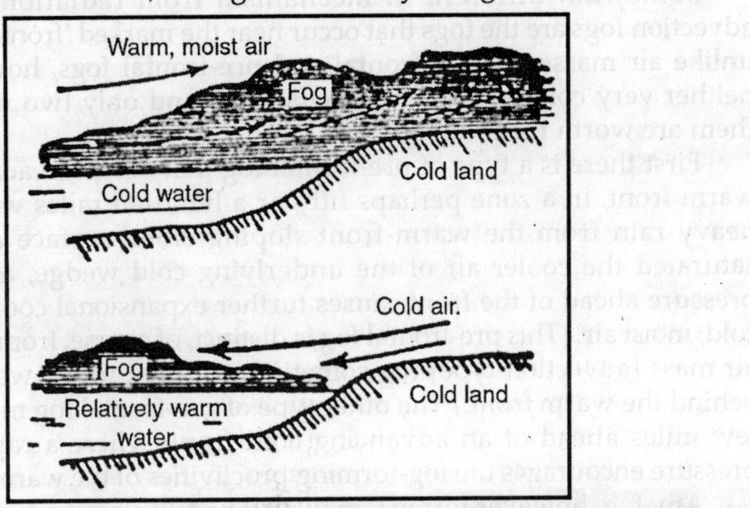

Fig. Advection Fogs

The second type of advection fog occurs when cold air flows over a moist, relatively warm surface such as ocean or lake water, or possibly land wet by recent rain or melting snow. Here the invisible water vapour, evaporating from the water surface, is chilled below its dewpoint by the cold air above so that it visibly 'steams.'

On land, the steaming of lakes on autumn mornings, as

radiation-cooled land air sweeps across water areas less prodigal with their summertime heat, shows the mechanism of cold-air advection fogs in miniature form. Similar steaming fogs occur in arctic regions when cold, belowzero polar air sweeps across leads of open water in the pack ice; though here, of course, the fog is formed of ice crystals.

Cold-air advection fog on the ocean is quite rare, but notably hazardous when it does occur. Near the United States it is confined mostly to the cast coast, and arises from the rapid southward and eastward outbreak of a large, cold (polarcontinental) air mass. When the fresh polar air, at near-zero temperature, rushes southward or southeastward behind a marked cold front over water at temperatures around 50° F. (or even 70° in the Gulf Stream), steaming fog arises on a widespread scale and perhaps masthead high. What with the viewless murk, the piercing cold, and a howling gale out of the north or northwest, even a steel steamship may do well to anchor inshore, despite every modern aid to navigation.

Frontal Fogs

Somewhat different in mechanism from radiation fogs and advection fogs are the fogs that occur near the marked 'fronts' between unlike air masses. These frontal and pre-frontal fogs, however, are neither very common nor very persistent, and only two varieties of them are worth mentioning here.

First there is a type of pre-frontal fog that occurs in advance of a warm front, in a zone perhaps fifty or a hundred miles wide where heavy rain from the warm-front sloping cloud-surface above has saturated the cooler air of the underlying cold wedge, and falling pressure ahead of the front causes further expansional cooling of the cold, moist air. (This pre-frontal fog is distinct, of course, from the warm air mass [advection-type] fog sometimes present in the warm sector behind the warm front.) The other type of pre-frontal fog may occur a few miles ahead of an advancing cold front, where a rapid fall in pressure encourages the fog-forming proclivities of the warm air mass.

Any fog, however formed, may drift bodily elsewhere and engulf regions well removed from its origin, provided aerological conditions do not favour its rapid evaporation. The front of the fog, particularly at sea, may take the well-defined form of a steeply sloping 'bank,' with a distinct and clear-cut edge. When a sea fog comes ashore, it remains on the ground if the land is colder than the water. If the land is relatively warmer, however, the lower surface of the fog lifts somewhat above the ground.

FORMATION OF CLOUDS

Fogs and clouds are of the same texture, and their different appearances are mainly a matter of viewpoint. High cloudturrets may gleam as white, as apparently solid, as alabaster in the bright upper sunlight; but explore the insides of those seeming-solid vapour-peaks in an airplane, and they become just as ragged, gray-murky, cold, and damp as any fog.

Having seen how cooling of humid air near the ground produces fog, we may well guess that cooling of some sort in the upper air produces cloud. How is this large-scale free-air cooling brought about? Not by conduction (for air is a very poor conductor of heat), not usually by radiation (for dry air neither absorbs nor radiates heat readily), but mainly by the expansion cooling of air forced in some way to rise.

In the preceding we examined the underlying mechanism of this rising-refrigeration process — we followed the salient facts of dry-adiabatic cooling, of moist-adiabatic cooling, of dry-air and moist-air stability and instability. It remains now to see how these various aerological conditions play their part in forming clouds out of thin air.

Forced Cloud Formation

If an entire layer of moist air is driven by prevailing winds up the slope of a mountain range, or drifts up the inclined plane of an underlying wedge of colder air, it cools throughout its whole extent at something like the dry adiabatic rate (about -5° F/1000 ft — or 5.5°, to be exact) so long as moisture remains entirely and invisibly vaporous. But at some moderate altitude, if its humidity is high, the rising air layer reaches its dewpoint.

Then moisture begins to condense on myriads of infinitesimal nuclei that are usually present in air — salt particles, haze particles, and so on — into the form of innumerable small water droplets visible in the aggregate as cloud. The individual droplets forming the cloud are about the same size as fog droplets or somewhat smaller — say 1/2500 of an inch in diameter. After this cloud condensation begins, of course, released latent heat retards the further cooling of the rising air to the moist adiabatic rate (say -3° F/1000 ft at moderate altitudes).

If the moist, rising air layer is ill-disposed towards updrafts and downdrafts within itself — if, in a word, it is stable — it forms merely one great, shapeless horizontal mass of flat, quiescent, layer-like cloud in the same way that a groundcooled air layer forms fog. A large proportion of all clouds are formed, as simply as fog, within such moist but stable air layers lifted in some way by external causes. If all clouds were so formed, the science of aerology would be much simplified.

But more complicated mechanisms of cloud formation produce some of the most splendid clouds and spectacular storms.

Free Cloud Formation

It is worth recalling from the stability in the preceding that, in 'absolutely unstable' air (lapse rate steeper than about -5° F/1000 ft), vertical currents tend to continue and to grow, once they are triggered off by commonplace causes such as surface heating, mountain updrafts, and so on. The same may be true of 'conditionally unstable' air (lapse rate between -5° and -2 1/2° F/1000 ft), which behaves like absolutely unstable air once condensation has begun.

By the 'beginning of condensation,' of course, we mean the formation of the minute water droplets that constitute visible cloud. The ultimate result of instability in any air layer, dry or moist, is updrafts (together with compensating downdrafts elsewhere). And the result of updrafts, in any moist air, is bound to be clouds.

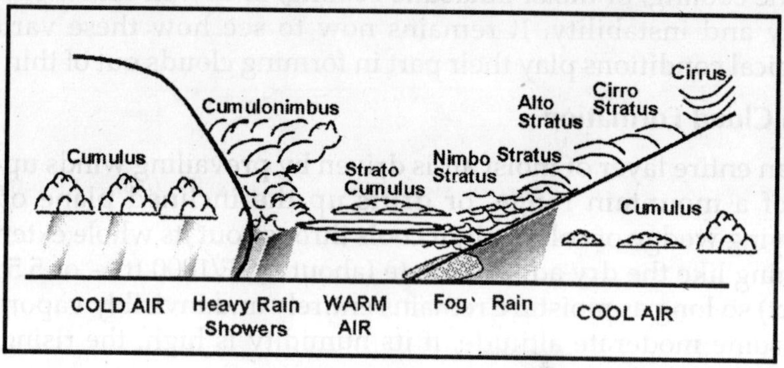

Fig. Cloud Formation

Unlike the flat, quiescent layer-like clouds formed by general air lifting under external forces, these cloud-children of instability are usually lumpy, heaped-up, puffy, and cauliflower-like in appearance. The instability clouds themselves may be very large, occasionally towering higher than the highest mountains, yet they tend towards localized lumps, turrets, and masses, called 'cumulus,' rather than widespread and shapeless layers, called 'stratus.'

In freely rising air, instability cloud-birth always begins at a certain definite level — the *condensation level* where falling temperature meets existing dewpoint. Under some conditions this condensation level is very definite, and readily apparent as a flat cloud base or 'ceiling.' The typical woolly, cream-puff clouds of fair-weather summer skies, for example, all have flat bases (and not the rounded bases that some artists like to paint into them); and these bases are actually all at the same elevation, though more distant clouds may appear lower to ground observ- ers by reason of perspective. Each detached, puffy cloud in this idealized summer picture represents, of course, the visible crown of an updraft (invisible below the condensation level) that is sustaining the cloud. In the clear spaces between these detached woolpacks there are usually downdrafts more gentle and more extensive than the updrafts that they balance. There is a rough but useful relationship between 'ceiling,' or height of cloud base, and surface temperature and dewpoint. At moderate temperatures, each degree of temperature-dewpoint difference means about two hundred and fifty feet of ceiling. Thus if the surface temperature is 70° F. and the surface dewpoint 66° F., the ceiling should be around one thousand feet.

Occasionally the mechanism of updraft-builds-cloud is revealed in detached and clearly apparent form by a large fire. The smoke from the fire rises, of course, along with a great column of heated air. Just as the smoke begins to thin out, perhaps, at some moderate elevation, a new and thicker turruted murkiness appears and builds upward. This upper murkiness is not smoke, but cloud formed by the updraft, and the elevation where it appears is the condensation level.

If a large and deep air layer is sufficiently unstable and sufficiently moist — conditions that exist in tropical or summertime thunderstorm weather — both the updraft and the consequent instability cloud may grow out of all proportion until the final result is vertical currents ballooning upward at possibly one or two hundred miles an hour, and a towering cloud mountain as high and wide as Everest.

For the wholesale condensation of water vapour into cloud droplets releases wholesale latent heat, which intensifies the updraft, which condenses still more moisture and releases still more latent heat, and so on. You can sometimes watch these thunderheads visibly growing against the blue sky — the faster they grow, the stronger the updraft and the more unstable the air to begin with. The upward momentum of rising air also helps in this cloud-building, which is ultimately checked, in extreme cases, by the great upper temperature inversion that is the stratosphere. Instability clouds are not limited to hot weather. It is all a matter of relative warmth and coldness alow and aloft — of lapsed rate, provided enough moisture for cloud-building is available. A strong wintertime outbreak of frigid polar air over a relatively warm water surface (say along the Atlantic coast of the United States, or across the Great Lakes) can produce as puffy and squally clouds as anyone could wish. And in winter, with the sun low in the sky, these instability clouds and snow squalls look blacker and more forbidding than most summer storms.

Combined Cloud Formation

Not all clouds are distinctly of layer-like, quiescent stratus type or indisputably of puffy, instability cumulus type. It is possible to have both types present in the same section of sky, perhaps at different levels. Hybrid or combination types, such as strato-cumulus, are also possible.

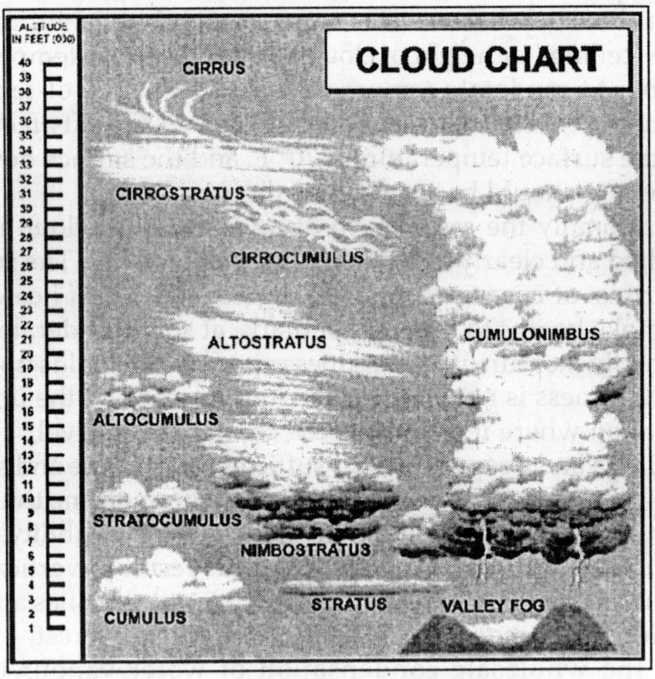

Fig. Clouds by Type and Altitude

The kind and prodigality of cloud-building depends on the relative strength and interplay of the three great cloud-building forces — moisture, instability, and external lifting. Any moist air layer bordering on instability tends to become more unstable, the more it is lifted by external forces. Such a layer, externally lifted, may at first form quiescent stratus clouds which, as the lifting goes on, may bulge upward here and there into growing instability cumulus turrets. Nevertheless, every cloud that we see in the sky must result from some chance combination of the three cloud-building forces. The more unstable the air is, the more moist it is, and the more violently it is lifted by external forces, the greater will be the bad-weather effects in clouds, rain, snow, hail, and the like. We have said earlier that radiation is usually an unimportant factor in cloud formation. But moist air absorbs and radiates heat better than dry air, and it is possible for a high, moist layer (having a relative humidity close to one hundred per cent) to radiate away enough heat at night so that it forms a layer of high stratus cloud. Possible — but not common. This ever-changing interplay of cloud-building forces in Nature's air ocean produces what, at first sight, seems to be an endless and bewildering variety of cloud forms. But these cloud forms (or the more common ones, at any rate) can be grouped into a few types and classes that are quite easy to remember and identify. The most basic cloud distinction of all, between layer-like stratus and heaped-up cumulus, has indeed already been made.

PARADE OF THE CLOUDS

The first attempt at naming clouds systematically and completely was made by Luke Howard, an Englishman, in 1803. He classified all clouds by shape or appearance into three general types — cirrus or 'curl' clouds, cumulus or 'heap' clouds, and stratus or 'layer' clouds. Other types of clouds, such as nimbus or rain clouds, Howard associated in some way with his three basic types.

During the nineteenth century other cloud classifications were proposed, but Howard's scheme, from time to time elaborated and improved, gradually grew into a standard system increasingly accepted throughout the world. This international cloud classification is explained fully and authoritatively in the *'International Atlas of Clouds and of States of the Sky,'* * published at Paris in 1932. The complete classification includes many elaborations such as 'altostratus translucidus' and 'cumulonimbus capillatus' — most of which have no great practical importance. In this book we are concerned only with the commoner clouds, such as any amateur can learn to name and recognize. And most amateur difficulties in the past have come, I believe, from too-exhaustive or illogical systems of classification. So we shall keep our cloud charts very simple, and as logical as possible. The most fundamental cloud differentiation, already hinted at in the preceding section, is based, like Howard's, on shape or form:

CLOUD CLASSIFICATION BY FORM OR SHAPES

Two Fundamental Shop, Classes

Layer class (from stable air)	Stratus (= 'a spreading out')	A quiescent cloud layer, rather flat and formless, extending horizontally for the most part.
Puffy class (from unstable air)	Cumulus (='a heap')	A piled-up heap of cloud, usually having definite three-dimensional shape, often growing vertically.

Two Special Shape Types

Wispy type	Cirrus (= 'curl')	White, fibrous mares' tails or feathery curls at very high altitude.
Broken type	Such as Fracto-stratus * Fracto-cumulus	Heat-evaporated or wind-torn into separate smaller cloud masses with ragged, wispy edges.

Mixed and Storm Typer

Combination type	Strato-cumulus	Combining some charicteris-tics of two different classes.
Precipitation types	Nimbus (= 'rainstorm') Nimbo-stratus Cumulo-nimbus	Clouds of stratiform or cumu-loform shape, respectively, from which rain, snow, or ice is actually failing.

*Hyphens in cloud names have been dropped by professional serologists, but they are retained here for the sake of simplicity and clearness.

Another way of classifying clouds, and perhaps a better one for understanding how they fit into the pageant of tropospheric weather, depends on their height above the earth's surface -or, in a more general sense, their altitude above sea level. Beginning with the highest and thinnest clouds, and ending with the lowest and thickest, the more common forms are:

Table. Cloud classification by altitude

Upper or citrus-like types	Cirrus Cirro-stratus Cirro-cumulus	Usually occur in the upper trop-osphere, often not far below the stratosphere, at heights of about 4 to 7 miles (6-10 km.).
Middle or high ('alto') types	Alto-stratus Alto-cumulus	Usually occur in the middle troposphere at heights of 2 to 4 miles (3-6 km.).
Lower types	Stratus Cumulus Strato-cumulus Nimbo-stratus	Usually occur in the lower troposphere, from the surface up to a height of nearly 2 miles (3 km.).
Vertical growth type	Cumulo-nimbus (thunderstorm)	Single cloud mass may extend from near the surface up to heights of 4 to 7 miles.

This table shows at a glance how various types of clouds fit into the general picture of the troposphere. From it one can appreciate why towering cumulo-nimbus clouds can be seen hundreds of miles away in clear air; one can sense why a spreading high deck of cirro-cumulus stands out in the red light of a clear sunset for perhaps three hundred miles to westward, where the sun-rays, receding at ten miles per minute, still light the cloud deck's edges half an hour after they have left the ceiling directly overhead. These high decks of upper clouds (cirro-cumulus or alto-cumulus), together with clear lower air to westward, are the one unfailing recipe for sunset splendor — were long before Meredith, and always will be:

'Ay, but shows the southwest a ripple-feathered bosom, Blown to silver while the clouds are shaken and ascend,

Scaling the mid-heavens as they stream, there comes a sunset
Rich, deep like love in beauty without end.'

What do all these cloud forms really look like, and how do they arise out of the vast emptiness of the air ocean? The cloud photographs and diagrams in this book offer a good part of the a nswer. And in further elaboration, we review briefly the various outstanding peculiarities of each cloud form in turn:

CIRRUS

True cirrus clouds are fibrous and delicate in appearance, showing as white wisps against the upper blue. Sometimes they arc combed out into mares', tails; sometimes they are formed more thickly into curls on feathery plumes or parallel 'plowed-field' bands. But always they are the most tenuous clouds of the sky, scarcely dimming the sun's disk, and the highest, floating at five to seven miles above the earth (8 to 11 kilometers).

As the temperatures at these levels are always far below zero, cirrus clouds are formed from minute, free-floating crystals of ice (simple hexagonal prisms). In fair weather they may result from general lifting of the upper troposphere above the active, 'cold-dome' portion of a high-pressure area. As foul-weather harbingers they are 'locks of the approaching storm,' blown out by substratospheric gales far in advance of the general cloud area of a cyclonic warm front.

Sometimes broken or wispy clouds fracto-stratus or (fractocumulus) in the lower levels have a superficial resemblance to cirrus clouds, and are mistaken for them by laymen. But the true cirrus clouds, carefully watched, are very evidently much higher than these fracto-forms.

Cirrus clouds are always interesting to watch because they show the motion of air near the top of the troposphere. Sometimes the long, fibrous tails seem to radiate from a cyclone centre below the horizon, and actually are combed out along the direction of air motion by the upper winds. At other times cirrus plumes or rays can be seen to move, not along their length but across it, like successive ranks of marching soldiers.

Here the cirrus plumes make visible a colossal wave motion in the substratosphere — white crests of upsurging condensation and blue troughs of downdropping evaporation — great, silent aerial surges that may measure ten or twenty miles from crest to crest. Some cirrus clouds, again, are probably nothing more than evaporating wisps of what was originally cirro-stratus -a sort of 'fracto-cirro-stratus,' one might say. After sunset or before sunrise, cirrus clouds often glow with beautiful tints of flame-yellow and rose-pink.

Cirro-Stratus

Cirrus-like cloud smoothed into one continuous, thin layer, resembling a whitish veil, is called ciffo-stratus. It may be smooth and milky, or again uneven and somewhat fibrous, forming out of thickening cirrus. Cirro-stratus never casts shadows, nor does it greatly dim either the sun or the moon, though in thickening and lowering form (approaching alto-stratus) it begins to blur their disks somewhat.

Cirro-stratus is almost as high as cirrus (3½ to 7 miles, or 5½ to 11 kilometers), and is likewise formed of minute hexagonal ice crystals. In all citrus-like cloud-forms, but most noticeably in cirro-stratus, these ice crystals may refract and reflect light in such a way as to form halos or luminous rings around the sun or the moon.

The most common halo is a single ring of about twenty-two degrees radius, colored faintly if at all; but one or even two additional outer rings may be seen occasionally, together with rarer 'mock suns,' 'sun pillars,' tangent arcs, and the like. ('Coronas,' or rare, colored sun-andmoon rings that may superficially resemble halos, are usually caused by water-droplet clouds at much lower altitudes.)

Cirro-stratus cloud is most frequently formed in the upper and outer warm-front portions of a cyclonic cloud area, thickening to alto-stratus towards the storm centre, thinning out to true cirrus wisps in advance. Hence while quiescent detached cirrus is a frequent fair-weather cloud, cirrus thickening into a cirro-stratus veil is a more definite portent of possible cyclonic approach. The portent is still more definite if cirro-stratus thickens to alto-stratus — but by that time, the rain is not far behind.

Cirro-Cumulus

Cirro-cumulus clouds are small, fleecy, white flakes or very small cloud balls — a 'mackerel sky' — perhaps merging into cirrus or cirro-stratus, and composed of supercooled droplets or ice crystals floating at great heights (3½ to 5½ miles or 5½ to 9 kilometers).

True cirro-cumulus clouds are not very common, and are easily confused with thinning forms of alto-cumulus — which latter are always well below the true citrus-like cloud levels. The distinguishing feature is, however, that in cirro-cumulus the individual flakes usually appear to be smaller by reason of their great height, and they are never thick enough to cast shadows or show dark undersides.

Cirro-cumulus clouds may sometimes be formed by small up-and-down currents along a wide but shallow unstable layer in the upper air. But more often they are arranged, like some types of cirrus, in parallel rows or lines, showing wave-motion between overrunning and underrunning air currents somewhat similar to ocean waves along the sea surface where water meets differently moving air.

All these high, cirriform clouds can best be seen, incidentally, through dark (preferably purplish) glasses; and they can best be photographed, like most other clouds, through red filters.

Alto-Stratus

This cloud usually takes the form of a thick grayish or bluish veil, always somewhat fibrous in texture. It is thicker and less translucent than cirro-stratus, which often merges gradually into it with the approach of a cyclonic storm. Through typical alto-stratus the sun and the moon gleam only weakly and vaguely, as through ground glass; they have a watery and diffused look, or may be almost wholly obscured, usually casting no shadows. Unlike cirriform clouds, altostratus is composed mostly of minute water droplets plus ice crystals scattered here and there.

The usual height of alto-stratus is 2 to 3½ miles (3 to 6 kilometers). It occurs most frequently along the slightly sloping surface of a warm front, where the overrunning warmer, moister air, ascending the inclined plane of the cold-air wedge, has lost by condensation a good deal of its original warmth and moisture. This frontal alto-stratus often marks a zone where light rain or snow is just beginning to fall from the high clouds, though this outermost precipitation (except cold-weather snow) may evaporate again before reaching the ground. Aside from the typical warm-front cloud-sheet, local alto-stratus of a more level and quiescent type can occasionally result from the radiational cooling of a high and very humid air layer.

Alto-Cumulus

Alto-cumulus is quite similar in general appearance to cirro-cumulus, but the individual flakes or cloud-balls of alto-cumulus are always larger and grayer. Usually they cast shadows and show dark undersides. The lower the alto-cumulus is, the larger are its opaque lumps. Certain forms of altocumulus ('castellatus') display little cloud-turrets growing upward from the main moisture-supplying layer.

In common with alto-stratus and most of the lower clouds, alto-cumulus is formed of minute water droplets, which may, however, be sub-cooled well below freezing in the liquid state. The usual height of alto-cumulus is 2 to 3 miles (3 to 5 kilometers). Like cirro-cumulus, alto-cumulus is often arranged in parallel rows along a wide and level deck, resulting from wave motion between differently moving layers of air. Alto-cumulus usually results from upper-air wave motion of some sort, or instability; it may grow into alto-stratus or nimbo-stratus.

Stratus

Any flat, shapeless, more or less uniform layer of cloud in the lower levels, usually dull gray in colour when viewed from below and whitish when viewed from above, is called stratus. Stratus resting on the ground is simply fog, formed and dissipated in the same way as fog; and high fog is simply low stratus. More usually, however, stratus occurs at heights of 1500 to 3000 feet (500 to 1000 meters). Low stratus is often a local cloud, gradually breaking away into shreds and smaller masses called fracto-stratus or 'scud.' Stratus can produce a fine drizzle, but not appreciable rain or snow.

CUMULUS

Cumulus clouds are thick, dense, detached 'woolpacks' or 'puffballs,' usually as high as they are wide. Their bases, as we have seen, are usually flat and uniformly fixed at the condensation level, but their upper parts are dome-shaped and variously bulbed and rounded, like cauliflowers. Their colour ranges through every nuance of sunlight and shadow from dazzling white above to dark gray below. It will bear repeating that a detached cumulus cloud always marks the top of a rising air current, which perhaps originated in the first place from an overheated portion of the earth's surface, or from some updraft in the general air flow. In any growing cumulus cloud conditional instability is being realised and condensation of moisture is contributing its latent heat to make the updrafts more violent. If moisture is plentiful and the instability extends upward to great heights, the cumulus grows into a cumulo-nimbus, or thunderstorm.

But if an ordinary growing cumulus bumps into a very stable layer such as a deep temperature inversion, its top flattens and its growth stops.

Most ordinary fair-weather cumulus clouds, formed under clear skies at levels between 2000 and 8000 feet (600 and 2500 meters), wax and wane daily with the sun (or with surface temperature). Foul-weather cumulus clouds, on the other hand, occur at any time of day along cold fronts and within moist, unstable air masses. A torn, immature, or partly evaporated cumulus is sometimes called fracto-cumulus. Sometimes cumulus clouds on the lee side of a mountain range evaporate in the dynamically heated downdraft so rapidly that they seem to disappear as if by magic.

Strato-Cumulus

Strato-cumulus usually occurs in the form of a wide and fairly level layer marked by thick rolls and dark, rounded masses of cloud. It somewhat resembles alto-cumulus, but its cloud masses are usually larger, thicker, and darker. Sometimes there are small or large open breaks between the rolls and masses; sometimes the layer is continuous but steeply undulating, especially along its upper surface as seen from an airplane.

Strato-cumulus clouds usually occur in the same levels as fair-weather cumulus — say 3000 to 8000 feet (10002500 meters) — and may be formed by the flattening of cumulus clouds, instability in a stratus layer, or wave motion between two differently flowing layers of air. Strato-cumulus can change into, or develop from, stratus.

Nimbo-stratus. This cloud type, formerly called plain 'nimbus,' is in general any low, thick, formless, dark-gray cloud layer from which rain, snow or other precipitation is actually falling. Even if the rain momentarily stops, this cloud has a ragged 'rainy' look, and is still called 'nimbo-stratus.' Below the main cloud deck there may be various ragged 'scud' clouds (fracto-stratus or fracto-cumulus) that occasionally merge into nearly solid lower layers or into the main cloud deck above. Nimbo-stratus transmits a feeble illumination that seems to come from within the cloud itself.

Usually nimbo-stratus develops out of alto-stratus and rainformed lower fracto-stratus or 'scud,' marking the near approach of a warm front in the central region of a cyclone. Less frequently, nimbo-stratus may develop out of heavy and threatening strato-cumulus. In either case the nimbo-stratus cloud, at heights of perhaps 1000 to 5000 feet (300 to 1500 meters), results from widespread condensation in air forced to rise by reason of its movement around or towards the centre of a cyclone. Scattered areas of nimbo-stratus may develop from cumulonimbus, as in the rear parts of a thunderstorm.

Cumulo-Nimbus

This is a great cumulus cloud mass in which condensation of moisture, release of latent heat, and increasing updrafts (fed by a deep layer of moist and unstable air) have grown to gigantic proportions. Its colour varies all the way from alabaster whiteness on the upper turrets to dark gray, black, or dark greenish below. The whole cloud mountain may tower from some ordinary cumulus level, such as half a mile, up to a height of five, six or even seven miles.

In a broad sense any large cumulus mass producing a shower can be called cumulonimbus, and this type is the common shower cloud, with or without thunder, of the tropics. But in temperate latitudes, cumulo-nimbus connotes more specifically a large, towering thunderstorm cloud whose top, blown out into anvil shape and cirrus-like texture by substratospheric winds, consists mostly of ice crystals. Hoods of alto-stratus may mark temperature inversions that the growing cloud mountain breaks through.

Special Cloud Forms

One special cloud form worth mentioning is mammato-cumulus or 'rain balls,' whose cumulus heads bulge downward instead of upward from a high cloud deck. It usually appears as a multitude of small, dark downward bulges in a turbulent layer of alto-stratus, alto-cumulus, or stratocumulus. The pendulous cloud pocks are evidently caused by local cold downdrafts perhaps started by detached showers of falling hail or snow. Mammato-cumulus always indicates very disturbed and unstable conditions in the upper air, seldom occurs far from a thunderstorm, and sometimes precedes tornadoes.

Another interesting cloud form is the detached, lens-shaped ovals that are sometimes formed at considerable heights out of quiescent or slowly dissolving cloud masses (alto-stratus 'lenticalus,' and so on). Sometimes the crest cloud on a mountain summit, caused by the steady flow of air up over the peak, takes this lenticular shape. The lens shape is produced, in any case, by symmetrical air flow and selective evaporation.

When high winds are blowing in mountain country, any tall detached peak is likely to fly a 'banner cloud' that streams out, pennant-like, far to leeward. Sometimes the great plume is formed of blowing snow from the peak itself. But if the humidity is just right, a true banner cloud can form on a perfectly bare peak such as Yosemite's Half Dome in summer, caused by the dynamic cooling effect of lowered pressure in the lee of the peak.

RAIN AND SNOW AND ICE

At ordinary temperatures all lower clouds are composed of minute water droplets perhaps 1/2500 of an inch across. Left to themselves in still air, these cloud droplets would sink slowly, at a speed less than one mile per hour. Yet as the droplets jostled together, somewhat larger droplets might be formed here and there. From a cloud mass (low stratus or high fog) near the ground in very moist air, these larger droplets, perhaps ten times as large as the original cloud droplets, might sprinkle downward in the form of a gentle drizzle.

But such a drizzle does not even approach the wetting power of real rain. Ordinary raindrops are ten times as big again as the drizzle droplets — say 1/25 of an inch across. And rain clouds, considerably less common than ordinary clouds, are distinguished from them by two characteristics: fairly strong rising currents, and considerable depth of cloud. Many rain clouds indeed probably extend, summer or winter, well above the freezing level, and ice particles probably play an important part in their mechanism.

In any case the cloud droplets of any active rain cloud are being borne steadily upward, though not quite as fast as the air surrounding them. In the upper parts of the cloud the droplets are probably farther apart, and somewhat larger by reason of longer and freer condensation. Here and there they are combining into still larger drizzle droplets. Well up in the clouds the rising currents, not very uniform anyway, must somewhere begin to diminish.

As a result the larger droplets begin to fall downward, or at least rise slower than most of the cloud droplets, and in either case begin to sweep through the smaller droplets and the moist air surrounding them, gathering unto themselves more and more moisture as they fall. Finally these growing drops perhaps become large enough to continue earthward, even through moderate rising currents, in the form of rain. Much tropical rain and some temperate-zone rain probably originates in this simple way.

But in temperate and arctic latitudes probably most rain, and perhaps all heavy rain and certainly all snow, evolve originally out of ice particles. These ice particles can sublime directly out of water vapour onto sublimation nuclei (dust particles and the like, always present in the atmosphere) very easily at low temperatures around 5° F. (-15° C.) or colder; or, perhaps less frequently, they can be frozen out of ascending water (cloud) droplets that are likewise cooled well below freezing and jostled sufficiently by turbulence. The level at which the ice particles are first formed varies with latitude and the season and the general weather situation — from as low as a mile or less in winter to as high as four miles or more in summer. The upper parts of a typical heavy rainstorm, therefore, are mostly iceparticle clouds — cirro-stratus, alto-stratus, and the like -rather thin-textured clouds with indistinct edges. The lower parts of the same typical rainstorm, at least during its earlier phases, are mostly water-droplet clouds — cumulus, stratus, and the like — opaque-looking and well defined.

As the ice particles from the upper clouds fall slowly (at perhaps two or three miles an hour) through the lower (waterdroplet) clouds, a curious transfer of moisture occurs. By reason of physical laws that need not be elaborated here, at certain temperatures and humidities water vapour is evaporated from liquid droplets, yet sublimed on solid particles.

Thus the falling ice particles constantly gain in bulk at the expense of the water droplets around them, and with increasing size fall gradually faster. And in addition to their sublimation-growth, they freeze to themselves any subcooled water droplets that they happen to hit. The ultimate form of the ice crystals depends on the humidity, and varies from simple ice needles to skeleton stars (snow) to snow balls (graupel). As these ice crystals fall, the ice-particle cloud itself thickens and lowers, perhaps encroaching on the water-droplet clouds below — or possibly engulfing them entirely, as in cold-weather nimbo-stratus. But after falling below the freezing level (provided it is well up in the air), the ice crystals soon melt into raindrops.

This is the idealized picture. But in a violent rainstorm -and most particularly, in a turbulent thunderstorm — cloud droplets, ice particles, raindrops, and snowflakes may all be milling around in some parts of the cloud.

RAIN

In 'light' rain the drops are very small, averaging perhaps 1/50 of an inch in diameter. Ordinary or 'moderate' raindrops are about twice as big (1/25 inch). Drops of 'heavy' rain are larger still, perhaps 1/16 inch wide. The very largest raindrops in a cloudburst may 1/8 inch or more across. Raindrops splashing on the ground of course look much bigger than the effective spherical sizes just given, and even while falling through the air they are somewhat flattened out. The largest possible raindrop size is about \! inch, which falls through still air at about twenty miles per hour; any larger drop, falling faster, would be broken into smaller drops by the wind of its fall.

Though all rain results from rising moist air, the gentleness and uniformity or violence and variability of the formative updrafts go far to determine what form the precipitation will take. Rain out of stable (stratus) cloud layers, gently rising, say, along a warm front, is light, steady, and continuous. But rain out of unstable cumulus along a cold front (where updrafts are patchy but powerful) is likely to be heavy, variable, and showery.

In the heaviest cloudbursts rain falls at the rate of three or four inches of solid water per hour. (This excessive rainfall rate cannot continue for long, but even an hour of it may cause floods.) A fall of nine or ten inches in a few hours, by no means rare, brings creeks and rivers raging out of their banks and inundates the countryside. The heaviest rainfall so far ever recorded was at Baguio, Philippine Islands, where forty-six inches — nearly four feet of solid water on a level surface — fell in twenty-four hours!

Rain is most beautiful when it falls through clear lower air in the near-level rays of a low sun, glistening and sparkling in all directions, and perhaps painting the spectral arch of the rainbow. The primary bow, and perhaps a secondary are, are always centered about the observer's shadow. Hence they can never form more than half a complete circle for anyone on level ground or at sea.

The cause of the rainbow is selective reflection and refraction of the various colors or wavelengths composing white light (at certain critical angles that vary slightly with the wavelength) by the small crystal-clear spheres of the falling raindrops. The primary bow, ranging from a sharp outer edge of red to a diffused band of inner blue or violet, has a radius of about forty degrees from the observer's shadow. The rarer secondary bow, if visible at all, begins rather sharply with red about nine degrees outside the primary and shades away outward into a nebulous greenish or bluish glow.

The larger the raindrops and the clearer the lower air, the more striking and beautiful are the rainbow colors. Ground observers, of course, see the largest bows when the sun is low. But from a high mountain-peak, or better still from an airplane flying between or below showery clouds, it is sometimes possible under a high sun to see a strangely unfamiliar but doubly miraculous and brilliant rainbow rounding into more than half a circle and seemingly laid on its side.

Less spectacular and less rare are the faintly colored aureoles, called 'glories' or 'mountain-bows,' that anyone on a sharp peak or in an airplane can see around his shadow on the white surface of a sunlit cloudbank. If the cloud droplets are near at hand, these glories may resemble faint but orthodox rainbows. Smaller aureoles that appear on more distant cloud-floors — faint colour-rings only two degrees or so in diameter — are probably due to selective diffraction of light rays around the sharp opaque edge of the peak or airplane.

SNOW

Few places on the earth are not visited by rain, occasional or frequent, light or heavy, steady or showery. Over the whole earth, indeed, it has been estimated that rain is always falling at the stupendous rate of sixteen million tons per second! But in northern latitudes, through a large part of the year, this never-ending deluge takes the form of feathery-white showers of snow — snow that, unlike rain, accumulates everywhere, in the valleys and on the hills, to store moisture against a future melting.

This snow blanket brings to sight general whiteness, and to sound general silence. Snow only four inches deep absorbs something like nine tenths of ordinary sound energy, being more efficient as a noise muffler than heavy draperies of velvet.

A feather-like snowflake floating slowly downward seems to have little in common with a piece of ice. Nevertheless, that is what the snowflake really is, a piece of ice usually crystallized into a very intricate and beautiful pattern. Snowflake patterns are so elaborate, consisting mainly of outgrowths at sixty-degree angles, that each flake presents a large air-catching surface compared with its weight, and falls slowly like a parachute. Snowflakes can be caught on a dark cloth and examined under a magnifying glass, and they have often been photographed with microscopic cameras.

Snow, as already indicated, is probably formed by the same general causative factors as rain, operating however at temperatures entirely below freezing. As the nuclear ice crystals fall, they freeze to themselves as crystalline extensions any subcooled minute water droplets in their path, and grow also by direct sublimation out of invisible water vapour. If snowflakes fall into temperatures above freezing on their downward path, however, each elaborate crystal melts into a drop of rain. In a climbing airplane it is not uncommon to sound through a region of rain up into a zone of snow (though the exploration is not particularly to be recommended, owing to icing hazards), and much winter rain undoubtedly begins as snow in the higher levels.

Although snow can melt into rain, rain cannot freeze back into snow. If rain from an upper, warmer air layer passes through a colder-than-freezing layer close to the ground, the result is frozen raindrops, or round pellets of clear ice, properly called *sleet*. Sometimes, however, snow falls in the form of round, compacted pellets which resemble sleet, but are more correctly called *graupel*; and occasionally these pellets come through to the ground. The remaining form of aerial ice is *hail* -pellets formed by repeated up-and-down rides in the towering thunderstorm clouds.

If rain falls in liquid drops on ground, houses, trees, or wires that have been previously cooled below freezing, it forms a layer of clear ice called *glaze*. 'Icing' of a somewhat similar sort can occur (to the embarrassment of pilots) on the wings or other parts of an airplane in flight through cold clouds or rain. Another type of ground icing (also noticed on airplanes) is *rime*, where subcooled fog or cloud droplets freeze on exposed projections, gradually building out into the wind. Widespread and continued glaze, which may bring down trees limbs, telephone wires, and power lines by the hundreds, is called an 'ice storm.'

The formation of clouds out of thin air, the birth of precipitation out of clouds — both these processes are physical experiments conducted by Nature in her open-air laboratory on a gigantic scale. Even a local summer shower involves millions of tons of air, hundreds of thousands of tons of water, and energy equivalent to the burning of thousands of tons of coal. A widespread storm, such as occurs along the path of a cyclone, may involve materials and forces many times greater.

Chapter 6

Wind and Climate Belts

This great, rotating ball that is Earth harbors within its atmosphere a thousand varieties of wind and weather over the diverse oceans and continents of its expansive surface, even at any one instant of time; and all these weather vagaries are constantly changing, quickly or slowly, from day to day, week to week, month to month.

Nevertheless there is a grand and logical scheme underlying general world-wide weather — a pattern that unmistakably appears in the average weather, *or climate*, experienced at representative places spaced up and down the meridians of latitude; and in the *prevailing winds* that arise out of these belt-like, world-encircling regions of climatic unlikeness.

This *average* planetary wind-and-climate picture is seldom, if ever, duplicated by actual world-wide weather on any single day — particularly in north-temperate regions, where actual weather variability seems to discredit the existence of any such regular scheme. Yet over the oceans — where the sea surface is relatively flat and relatively frictionless to wandering breezes, where sameness of temperature discourages climatic oddities and local weather freaks, where the climate belts and prevailing winds are most evident, typical and regular — it is possible to glimpse in actual weather some prominent features of the grand planetary pattern.

You can best sense certain of its gradual but definite and continuing changes, from latitude to latitude, by cruising north or south in a seagoing ship. Here, for example, is the actual record of a March cruise from Panama to Philadelphia in the old but able American freighter *Timber Rush* — a slow two-thousandmile cruise from torrid to temperate latitudes at the comfortable ocean-going dog-trot of ten knots:

> ***March 11th***: latitude 10° N, off Panama coast, temperature 79° F, broken clouds, wind ENE moderate (trade wind), caught barracuda on trolling line.
>
> ***12th:*** lat 12°, in Caribbean, temp 79°, scattered to broken clouds,

wind NE X N fresh to strong (norther), sea rough, water temperature 78°.

13th: lat 14½°, in Caribbean, temp 77°, scattered clouds, wind ENE fresh (intensified trade), water very rich blue.

14th: lat 18°, just southwest of Haiti, temp 78°, scattered, clouds to clear, wind E'ly light (in lee of Hispaniola).

15th: lat 22°, near Bahamas and Tropic of Cancer, temp 77°, scattered clouds, wind E'ly to variable light (horse latitudes) scattered clouds to clear.

16th: lat 26½°, in Atlantic, temp 73°, showers and thundershowers along a cold front, wind SW'ly moderate shifting to NW fresh with frontal passage.

17th: lat 30½°, in Atlantic, temp 67°, scattered to broken clouds, wind NW moderate (prevailing westerlies), water dark steely blue, water temperature 74°.

18th: lat 35°, off Cape Hatteras in Gulf Stream, temp 60°, broken clouds, wind N'ly fresh and increasing, water temperature 72°, passed school of whales.

19th: lat 38°, entering Delaware Bay, temp 35°, clear, calm to S'ly light, water temperature 50° after leaving Gulf Stream, 42° near coast, water dirty green.

Thus we passed, in the last four days of slow steaming at two hundred and forty easy sea miles a day, from a tropical climate to a temperate climate (which latter, in March, is mostly intemperate), from smiling summer skies and tufted cloudlets to cold, murky storms and driving showers, from moist, warm, enervating sameness to irritating and unexpected variability.

CLIMATES, TROPIC AND ARCTIC

Variability — that is the outstanding characteristic of temperate-zone weather. Even the brief sea log just quoted shows how weather in the tropics is much the same from place to place and day to day; temperate-zone weather, conversely, is diverse and changeable. But had our voyage extended far enough northward, it would have reached another region of comparative weather sameness — cold sameness instead of warm sameness -the arctic zone.

The temperate zone can have no stable and typical weather of its own. To southward (in the northern hemisphere) is the steady, warm, moist climate of the tropics. To northward is the fairly steady, mostly cold, fairly dry climate of the arctic. The temperate zone is a buffer region between tropic and arctic climates, a battleground on which the gigantic atmospheric forces of heat and moisture advance and retreat, an unstable compromise between irreconcilable extremes, swayed this way and that in its evanescent weather as tropic forces (warm air masses) or arctic forces (cold air masses) momentarily gain ascendancy.

Tropic Climate

Tropic wind-and-climate belts migrate somewhat, north and south, with the apparent north-south seasonal swing of the sun. To get within-the-tropics weather in its most typical form we must go well south of the Tropic of Cancer, nearly down to the equator — say to Panama in about lat. 8° N. In these latitudes there are only two main seasons — a winter dry season under the influence of the steady northeast trades, and a summer wet season brooded over by the fitful calms of the doldrums.

The outstanding features of tropical-marine climate are that the average temperature is practically constant from month to month (nor does it ever vary, by more than a few degrees, from day to day); and that the cloudiness and rainfall, always appreciable, reach high values during the wet-season doldrums invasion of summer and early fall.

Upper winds over Panama, at half-mile altitude, average northeast about 8 m.p.h. in the rainy season and north-northeast about 20 m.p.h. in the dry. At two miles altitude there is less seasonal change — the average wind only varies from east 10 m.p.h. (rainy) to cast-northeast 12 m.p.h. (dry).

Upper-air characteristics over the Caribbean Sea, quite typical of tropical-marine climate in general, were studied during the 1939 hurricane season by a United States Weather Bureau expedition to Swan Island. This four-mile-long islet, named after a nineteenth-century pirate, lies almost exactly in the middle of the Sea at about lat. 16° N., between Jamaica and British Honduras.

Within four months, observers Rhamlow and Leech released some two hundred and fifty pilot balloons and one hundred and twenty radio-sounding balloons. They found that the trade wind extended well aloft, though with diminishing force, here as in Panama — prevailing easterlies all the way up to about two and a half miles altitude. But above this level, the upper winds are mostly out of the west and increase with altitude.

Surface temperatures at Swan Island, and all over the Caribbean area, average about +75° to 80° F. the year round. But at the base of the stratosphere, ten miles up in these latitudes, it is cold indeed — about — 105° to — 110° F. Thus in the thick tropical troposphere there is an average lapse rate of something like — 3½° F/1000 ft (about — 6° C/

Km). Together with plenty of warmth and moisture in the lower levels, this lapse rate means that conditional instability must extend through considerable air layers, particularly in the vertical-sun-warmed rainy season of summer and fall — and it is this conditional instability, together with inexhaustible supplies of moisture from the warm ocean, that produces the daily torrential showers and thundershowers, the regular afternoon downpours and the moist, misty mornings which can make equatorial regions such a continual sweat-bath for ill-adapted white men.

On or close to the ocean this deep-tropical weather is bearable, or perhaps even comfortable. But the steaming jungles inland, where temperatures can rise unchecked and where all breeze is choked off by the tangled uprearing of dense vegetation, are sometimes aptly described as 'green hells.' As Julian Duguid wrote '. . . a rich eternal garment of green, dappled with golden sunspots . . . a dense, fever-stricken thicket shimmering in the heat with a perpetual glassy haze. . . .'

However, by no means all the land areas within the tropics are covered with jungle — some of them are sandy and arid in the extreme, and doubly hot though not so muggy. (The word 'jungle' itself, originally, meant only 'waste land' in India.) And even in the fecund jungle areas of the tropics, such as Panama and northern South America, vegetation waxes lush and green with the rainy season, wanes sparse and tawny with the dry season.

There are also irregular year-to-year variations in moisture and prevailing winds that change the face of tropical nature. I got off a ship on the Colombian coast in August, 1939, and walked inland for a while, expecting to wade through wetseason green tangles of vines and trees peopled with tropic creatures — but the soil proved to be dry and sandy, the jungle a dried-up shrinkage of scrub, the creatures hidden away from the glaring vertical sun.

Arctic Climate

Before Admiral Peary of the United States Navy marched five hundred icy miles on snowshoes from Ellesmere Land to the North Pole in the early spring of 1909, and since his time, many expeditions have entered the northern arctic regions. No latter-day explorer, incidentally, has ever surpassed Peary's record of intelligent planning and courageous, competent performance. But the airplane has rung startling changes in arctic travel.

In May, 1937, four large air transports set down four Russian scientists and a dog, together with a little portable cabin and food and fuel for a year or more, on one of the uncertain drifting ice floes that cover the Arctic Ocean around the pole. Leader Papanin, astronomer-aerologist Federov, hydrologist Shirsov, and radioman Krenkel had hoped to remain at or near the earth's axis for a whole year of observations, but their ten-foot-thick ice floe soon began to drift (as Peary or Fridtjof Nansen could have foretold), at first very slowly, across the two-mile-deep polar basin towards Greenland. Then it moved, less gradually, southward along the east coast of that high continent, and finally sailed at twenty-five miles a day on southward. The floe-campers were rescued by icebreakers and airplanes twelve hundred miles south of the pole in February, 1938.

Papanin's expedition found the summer temperatures of the polar basin hovering mostly around + 30° to 46° F.; clear spells of a day or two alternated with dull, overcast, rainy-to-wetsnowy weather. Winds varied from calm to light or moderate, mostly out of the north or northeast. Much of the ice was melting in the warmer spells, adding to the troubles and perils of the ice-campers. There was always the danger that their floe would disintegrate and dump them unceremoniously into the Arctic Ocean.

When winter came they had drifted several hundred miles south of the pole itself, but winter climate around the polar basin is probably pretty much the same everywhere. Temperatures ranged all the way from 50° F. below zero to 30° or more above, in accordance with the sweep of unlike air masses and cyclones, which occasionally extends well into the Arctic. For several months in midwinter the polar region is, of course, in continual darkness, with the sun visible at first in the spring only as a pale horizon glow around noon. On the thirty-first of December, strangely enough, the Russian wind-powered radio reported a temperature of + 43° F. over the arctic ice! But within a few hours, the mercury had tobogganed to a more usual value of -31° F.

The general north-polar region is, of course, about midway on the most direct (great-circle) air routes between Europe and Asia, and between eastern United States and Asia. Some years hence, perhaps, great 500-mile-an-hour strato-transports will be winging daily across these icy-watery wastes.

Compared with other climatic regions, the polar area offers good flying conditions — during at least two or three days out of a week, that is, and particularly at higher levels. Spring, when the polar high-pressure area lies symmetrically over the geographic pole, and not over the continental 'cold poles' in Greenland and northern Siberia, is the best flying season. A sealed-cabin, supercharged stratoliner, in any case, would be well up in the low arctic stratosphere (above thirty thousand feet) most of the time. Aside from Ellesmere Land at the north tip of America, one of the best fixed bases for observing the climate of the north-polar basin is the small island of Franz Joseph Land in lat. 81° N. long. 60° E., about six hundred land miles from the pole. In 1932-33 the Russian scientist Moltchanov (incidentally one of the pioneers in the development of radio-sounding balloons) made both surface and upper-air observations on Franz Joseph Land for a year.

Like the floe-campers, he found average summer surface temperatures around +30° to 40° F. Winter temperatures averaged around -10 F. Above the ground, for the first mile, there was usually a temperature inversion amounting to 5° F. or so; above that height the lapse rate or vertical temperature gradient was always surprisingly uniform at -3° or -4° F. per thousand feet up to the low polar 'tropopause,' marking the beginning of the stratosphere at about five miles altitude in winter, six miles in summer. Temperatures at the stratosphere base averaged about -75° F. in winter, about -40° F. in summer. July and August were the warmest months, as usual in the northern hemisphere; December was the coldest month at the surface, February coldest aloft.

The southern hemisphere in general, being mostly water, shows more typical wind-and-climate belts than the northern hemisphere, which is all cluttered up with land masses that get too warm in summer and too cold in winter. By the same token, or a reverse one, the antarctic climate, founded upon a high, icecovered continent symmetric around the geographic pole, is more typically 'polar' than the arctic climate. Antarctica is, of course, almost as well known to book-readers and moviegoers as Times Square — chiefly through the efforts of Admiral Byrd.

During the 1933-35 Byrd occupancy of 'Little America' in about lat. 79° S., long. 164° W., aerologists Haines and Grimminger recorded a great deal of antarctic weather. Mostly the average temperature was around +20° to 25° F. in summer (January), and around -35° to -40° in winter (July) — considerably colder, that is, than the north-polar climate. The coldest summer temperature was +12° on January 1; the warmest, +37° on December 22. In winter the mildest reading was +23° on July 27; the coldest was -71° on July 21.

During a roaring blizzard in July the temperature fell to -65° with the wind sixty miles an hour out of the southeast — and wind, rather than temperature, is what really hurts in cold climates. These southeasterly blizzards in Antarctica correspond to the northeasterly gales around the north-polar basin.

High winds in Antarctica sometimes come from the north, just as north-polar gales can arise out of the south, but the colder blizzards from somewhere around the South Pole are more usual. Laurence M. Gould, Byrd's geologist and deputy in 1928-30, writes of one terrific southerly gale in the Rockefeller Mountains that first blew away the heavily packed snow wall in front of his firmly anchored airplane, and then finally blew the plane itself away and wrecked it. Some time before the maximum wind came, the airplane's air-speed meter indicated a steady velocity of ninety miles per hour, with gusts well over a hundred![2]

Conquest of Climate

The decade from 1900 to 1910 signalized an. amazing conquest of the tropics as a livable environment for civilized man. The years after 1940, likewise, may see a parallel conquest of the arctics.

Within the tropics, except in certain areas plagued by extreme and wearisome heat, general living conditions were easy enough — too easy for man's own good, in fact; and aside from air transport, radio communication, mechanical refrigeration, and other new conveniences, the conquest was nearly all medical. Such dread warm-country diseases as typhoid, cholera, and yellow fever were finally vanquished — at least in the case of selected, well policed areas.

The work of Walter Reed, his assistant surgeons, and the heroic American soldiers who voluntarily entered the valley of 'yellow jack's' macabre shadow -all this really conquered the tropics and equatorial regions for white men, and contributed more than engineering and steam shovels to the triumphant completion, in 1913, of the world's greatest inter-ocean canal.

Within the arctics and around the poles, on the contrary, disease is unknown unless brought to the scene by human carriers, but living conditions (without many bothersome precautions) are unduly hard and exacting. Aside from the extremes of dark and cold and gale and blinding snow, the chief hazards are: difficulty in finding your way around (particularly in the long, cold, dark arctic night); difficulty in surviving ordinary mischances such as falling into water, badly tearing clothes, and so on, particularly if alone; difficulty in getting adequately balanced diet and sufficient ultra-violet radiation.

Yet even now, with our immense progress in transportation and communication, some of these difficulties can be largely overcome at a properly organized arctic station. Clothing must, of course, be of arctic type — layers of wool or fur inside for warmth, layers of canvas or leather outside for wind-breaking. Adequate shelter, plenty of fuel, and fresh food can now be transported easily by airplane to almost any arctic site. On location, howling arctic gales will provide plenty of electric power for lighting and radio communication, and perhaps for heating as well.

Many of the dark terrors of the outer arctic night, which have heretofore held men prisoned for months in their base camps (except in the full of the moon) can perhaps be conquered by such devices as more powerful and efficient portable searchlights (using super-hot filaments, say, or possibly fluorescent light-sources); portable radio-telephone sets for constant two-way communication with and direction-finding on one or more base stations; more compact and accurate instruments, and simpler methods, for celestial navigation.

Against the white, too-uniform snow-glare of the polar day, improved goggles (perhaps of polarizing type) may be employed. Lighter, stronger and warmer tents and clothing can perhaps be devised. Much of the heavy local leg-work of arctic exploration can now be done with snow-mobiles, either tread-driven or propeller-driven. By means of vitamin diets, ultra-violet sunlamps and radio-diathermy, human beings near the poles will perhaps be enabled to maintain a high state of health during the long arctic night, and to overcome quickly any respiratory diseases that do develop.

Eventually we may even see arctic health resorts, rest camps and play camps. Ellsmere Land in northern Canada (Peary's jumping-off place), less than three hours' flying time from the North Pole, is only twelve hours by present slow air liner from New York City. 'Polar Dude Camp on Ellsmere Land — spend next weekend here in the dark restfulness of the polar night under the flickering curtain of the glorious aurora borealis; pack ice exploring parties for the more active' — and so on. Such an advertisement may not startle our children twenty years hence.

TEMPERATE CLIMATE

We have said that the temperate zone is, in a climatic sense, nothing better than a battleground between tropic and arctic weather forces — between warm and cold air masses. Nevertheless the temperate zone (and in particular, the north-temperate zone) is that part of the earth in which civilized man evolved, and is probably the region most favorable to his continued progress and development.

On the *average*, of course, temperate temperatures are pleasantest and best for all human activities — neither too hot nor too cold — though this average may rarely be matched by the vagaries of daily weather. And even the extreme moodiness and changeability of temperate weather has its advantages. Variety in weather, at least as to the four- or five-day spells and sequences so common in middle latitudes, perhaps satisfies some deep, innate rhythmic urges in man which cannot be allowed to atrophy without intellectual and cultural and physical loss. However such obscure causes may or may not operate, it is certain that most human progress up from the dumb-ape level — most human art and ingenuity and imagination, most human science and engineering and social organization — has been best nurtured so far, and probably will be best nurtured in the future, in temperate latitudes.

Among all the various meteorological factors that go to make up a climate, temperature is the one that most directly influences human health, happiness, and efficiency. How do various temperatures really feel to average human beings? Different degrees of heat of course feel quite different to different people, or to the same people under different conditions — in particular, under varying conditions of exertion. But it is possible to indicate average human reactions to temperature, and they are somewhat as follows:

Table. Outdoor Temperatures feel to the average person

Temperature		Human Effects
(Fahrenheit)	(Centigrade)	
Below 0°	Below - 18°	Unpleasantly cold without heavy arctic-type parkas, etc., unless exertion is continuous and heavy.
0°-32°	- 18°- 0°	Feels cold with anything but arctic-type clothing, if exertion is light and exposure long.
32°-50°	0°-10°	Under same conditions, feels comfortable with heavy overcoat, hat, etc.; cold otherwise.
50°-65°	10°-18°	Feels cool with ordinary indoor clothing, comfortable with light overcoat, etc.
65°-75°	18°-23°	Feels comfortable with ordinary clothing.
75°-90°	23°-32°	Feels warm co hot if humidity is high, wind light; feels comfortable in bathing suit.
90°-100°	32°-40°	Feels hot under any conditions, unpleasantly hot with high humidity and light wind.
Above 100°	Above 40°	Unpleasantly hot under all conditions.

High humidity makes warm temperatures feel warmer, by retarding evaporation from the body. High humidity also makes cold temperatures feel colder, by making the clothing conduct heat better away from the body.

Wind makes cold temperatures feel colder, by removing the cushion of warm air that tends to form around the body in the absence of wind. This wind-cooling effect is particularly powerful at below-zero temperatures. On the other hand, wind makes warm temperatures up to 100° F. feel less warm, by increasing evaporation from the body. But if the temperature is over 105°, wind increases the discomfort.

Table. Indoor temperatures feel to the average person

Temperature		Human Effects
(Fahrenheit)	**(Centigrade)**	
Below 64°	Below 18°	Unpleasantly cool with ordinary clothing.
64°68°	18°-20°	Too cool for comfort if sitting still.
68°-70°	20°-21°	Optimum for health and comfort.
70°-74°	21°-23°	Warm but not uncomfortable if clothing is light.
74°-78°	23°-25°	Needlessly warm for healthy people.
Above 78°	Above 25°	Unpleasantly warm for healthy people.

The optimum temperature range, 68°-70° F., feels more comfortable if the air is normally moist.

For health and comfort, indoor relative humidity should be around sixty per cent. But in the average home or office, in winter, it is seldom more than twenty per cent. Evaporating pans on radiators help some, but not very much. Constant-duty electric evaporators, preferably combined with electric lights, might well be used in temperate-zone houses during the winter.

Today's temperature anywhere is indicated by the thermometer; tomorrow's (which in temperate or arctic latitudes may be shockingly different) is foretold by a good forecast. Beyond these limits, our best present guide to future temperatures is the 'normal.' To get the normal temperature for any particular day of the year at any particular place, the average temperatures for that day, observed over as long a span of years as possible, are all averaged together.

The normal temperature is, of course, merely a fictitious quantity seldom or never duplicated by any actual daily average; for climatically speaking, we dwellers in the temperate zone are always traveling now south, now north with the ebb and flow of warm and cold air masses. Is our North American temperate-zone climate changing? This question keeps recurring in weather talk, and most of the oft-quoted answers foist off something about 'colder winters in Grandfather's day' — a notion perhaps promoted by oldsters in support of their claims to general superiority. The rising generation, contrariwise, have been inclined to charge the whole idea off to dotage and general boastfulness. A definitive answer has been given by J. B. Kincer of the United States Weather Bureau:

For a good many years up through 1936 there has been a decided tendency to warmer and drier weather, the trends being especially marked during the past quarter of a century, notwithstanding an occasional bad flood or a severely cold winter. Take the winter season, for example: With the exception of those for 1917-18 and 1935-36, the winters for the past twenty-five years or more, considering the country as a whole, have been rather uniformly warmer than normal, and, on an annual basis, every year since 1929 has had above normal temperature.

Also, in the matter of rainfall, there has been an equally marked tendency to droughts in recent years. However, an examination of the longer weather records of the country, going back one hundred years or more, indicates that this does not represent a permanent change of climate, but rather a warm, dry phase of our normal climate, to be followed, doubtless, by a cooler, wetter phase, when there will be more rain in summer and lower temperatures in winter. . . .

Since the turn of the century, Mr. Kincer goes on to say, there has been a distinct yearly trend towards warmer winters and drier summers all over the world — and this in spite of a few exceptionally cold winters here and there. In other words, there are long (say fifty-year) cycles in climate, just as there are short (say five-day) cycles in weather. If all these cycles were only considerate enough to maintain constant length, amplitude, and form, both weather forecasting and climate forecasting would be child's play. But the form, intensity, and length of weather and climate changes are all complexly variable. In any case the semipermanent (say 500-year) picture of our American climate shows, apparently, no appreciable change. But the geologic-time picture, measured in thousands or millions of years, would again show marked climatic cycles — the advance and retreat of the various ice ages.

THE PLANETARY PATTERN OF WINDS

Just as there are on this great globe fairly definite zones of *climate*, so there are even more definite belts of *prevailing wind*. Some of these round-the-world prevailing winds are quite steady and uniform, such as the trades and anti-trades. Others, such as the prevailing westerlies of temperate latitudes, are extremely variable from day to day and week to week. But the *average* wind directions observed in temperate latitudes prove that here, too, there is a great round-the-world wind system from west to east — winds that blow generally across continents and oceans, across desert sands, sawtooth mountains, fields of rippling grain, waving trees in virgin forests, and wide expanses of heaving, restless, white-capped blue water.

The winds of the world are in essence a heat-pressure mechanism similar to that which causes a sea breeze, though enormously larger and more complex. The driving force behind all earthly winds, and all other earthly motions, is the enormous outpouring of radiant energy from the sun, our celestial ancestor and gigantic power-house ninetythree million miles out in space.

This radiant energy falls nearvertically on equatorial regions, and very obliquely on polar regions. Hence the equatorial regions are warmer, and are receiving more heat than they lose by earth-radiation. The poles, on the other hand, are radiating away more heat than they receive by sun-radiation. Hence, unless the tropics are to become unbearably hot and the arctics are to chill down to absolute zero (about -460° F. or -273° C.), there must be some atmospheric circulation between the two.

A heated region usually develops low pressure at the surface and high pressure aloft, that a cooled region develops high pressure at the surface and low pressure aloft. The result, where only short distances are involved, is that lower winds blow generally from cold to heat, and upper winds blow generally from heat to cold. Hence we might expect, as part of the planetary wind picture, some tendency towards an equatorward drift at the surface together with a poleward drift aloft.

However, that any largescale wind is gradually turned aside, to the right in the northern hemisphere and to the left in the southern, by the earth's slow rotation under the wind in question. In other words this 'earth's deflective force,' so called, tends to turn any equatorward-drifting air westward, and to turn any poleward-drifting air eastward.

In the upper troposphere, say at altitudes above two and a half miles, the *average* world wind pattern is a grandly simple thing that stems directly and understandably out of these elementary heat-pressure-wind principles. Upper pressure is relatively high over the equator, and relatively low over the poles, with a fairly uniform average horizontal pressure gradient between the two. Hence the average upper-tropospheric winds of the northern hemisphere, directly driven by this pressure gradient and swayed by the earth's deflective force, are from southerly to southwesterly to westerly (drifting, that is, generally towards the northeast, or spirally towards and around the pole).

In the southern hemisphere, driven by similar but oppositely directed forces, the average upper-tropospheric winds are mostly northwesterlies, drifting generally southeastward. Either of these average drifts can be temporarily upset by regional vagaries, such as a southward surge of cold upper air in the north hemisphere. But as a norm or average condition, the two great upper drifts perennially move from the equator spirally eastward and poleward in each hemisphere.

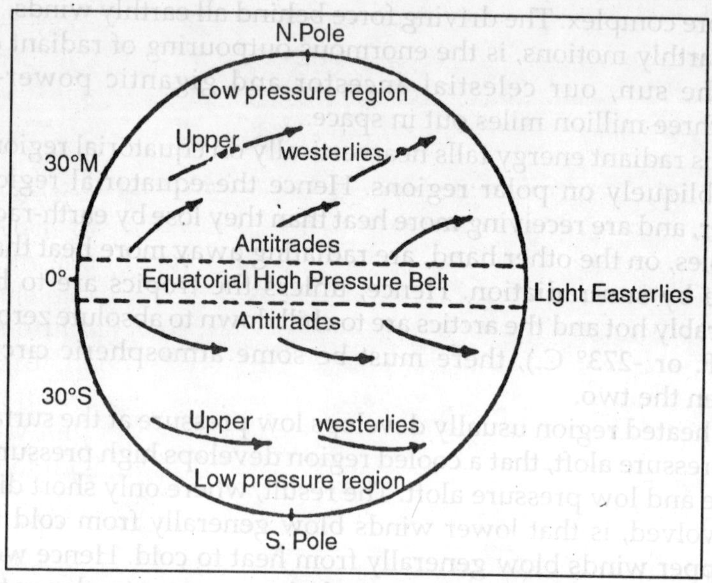

Fig. 41. Upper-Wind System of the World

The planetary pattern of lower winds (below the two-and-ahalf-mile level) is more complex, even in idealized form. For the lower winds are influenced by surface friction, and the earth's deflective force acts on them as a more drastic limitation on poleward or equatorward progress. Thus the planetary lowerwind pattern includes, not one wind belt in each hemisphere, but three. Moreover, these wind belts, and the pressure belts which separate them and regulate them, are in fact much modified and distorted (at least in the northern hemisphere) by the alternating surfaces of great continents and great oceans.

Perhaps the easiest and best introduction to them would be a voyage through ideally typical weather at just 5° of latitude (three hundred sea miles) each day down the whole 'length' of the world, along the ocean-parting meridian of long. 30° W. from arctic to antarctic. Any such imaginary meridianal voyage would cut across the following idealized or average wind-andweather belts: a zone of arctic easterlies above lat. 60° N.; a stormy 'polar frontal' zone (actually variable in both width and location) around lat. 60° N. or south of this; prevailing southwesterlies-to-westerlies (the familiar prevailing winds of temperate latitudes) between lat. 60° N. and lat. 30° N.; a dry, clear calm belt called the 'horse latitudes' (probably because horses often died here on becalmed sailing ships) around lat. 30° N.; northeast 'trades' (so called by reason of their steady, perennial character and consequent importance to commercial sailing ships) from south of lat. 30° N. down towards the equator: a humid, oppressive, and showery calm belt called the 'doldrums' (because of their dispiriting or dulling or 'dolting' effect on human sailors unwillingly becalmed in them) around the equator; southeast trades from near the equator down to somewhere near lat. 30° S.; the southern horse latitudes around lat. 30° S.; the southern prevailing westerlies between lat. 30° S. and lat. 60° S.; the southern polar-frontal zone around lat. 60° S. or north of here; and the antarctic easterlies south of lat. 60° S.

The steady poleward air transfer of the prevailing westerlies and southwesterlies in our latitudes is compensated by occasional large and violent southward outbreaks of polar air (cold waves), especially in winter, when the polar anti-cyclone extends down over Alaska and northern Canada. Southward outbreaks sometimes also occur far aloft, with 100-m.p.h. north winds in the stratosphere. Similar equatorward (northward) outbreaks occur in the southern hemisphere.

Considering first the northern hemisphere: There is an average high-pressure area, or anti-cyclone, around the 'cold pole' (despite the fact that many arctic observers have noted low pressures). This polar high-pressure area drives the cold, arctic lower air spirally outward (or southward and westward) as the arctic easterlies (mostly northeasterlies) down towards lat. 60° N. Along lat. 30° N. (the horse latitudes) there is a subtropical high-pressure belt, the northern slope of which drives the prevailing westerlies (mostly southwesterlies) up towards lat. 60° N.

Around lat. 60° N., or south of there, these two great and conflicting wind systems meet along the low-pressure belt of the polar-frontal zone — a stormy region which may actually swing far north or south, and may extend as separate and vari-formed fronts, between various unlike air masses, through a wide sweep of latitude. The subtropical high-pressure belt in lat. 30° N. (or its southern slope) also drives the northeasterly trades, as a general southwestward-to-westward drift around the earth, down into the equatorial low-pressure belt of the doldrums.

In the southern hemisphere, the subtropical high-pressure belt around lat. 30° S. drives the southeast trades up into the doldrums, and also drives the southern prevailing westerlies (mostly northwesterlies) down into the low-pressure polar-frontal zone around lat. 60° S. or north of there, where they meet and battle with the antarctic easterlies (mostly southeasterlies) from the south-polar high-pressure area.

In temperate latitudes, both north and south, lower winds have the same *general* or *average* direction as upper winds, and the prevailing westerlies form a deep *average* pattern (seldom fully duplicated in actual existing winds) that extends with increasing force up through the troposphere, moderating again in the lower stratosphere, and then mostly reversing, in light to moderate intensity, above twelve miles or so.

Within the tropics, on the 'winter side' of the equator, the easterly trades give way, at about two and a half miles altitude, to the planetary upper westerlies. Called the counter-trades or anti-trades, these tropical upper winds blow mostly from southwesterly in the northern hemisphere and northwesterly in the southern. The same planetary upper westerlies are present, to some extent, in arctic regions above the arctic easterlies.

CENTERS OF ACTION

The whole idealized wind-belt picture described above is based, it will bear repeating, on *average* conditions. Daily departures from this average, in any one region, may be very wide. It is also based on the conception of a *uniform earth* — either all water, or all uniform land. As a matter of actual fact the southern hemisphere is so nearly all water that this idealized pattern actually holds there much of the time; and the pattern may also hold, though less consistently, well out in the Atlantic and Pacific Oceans of the northern hemisphere.

But in general the northern hemisphere is something like equally divided into great land surfaces and great water surfaces, which vary greatly in the frictional drag they oppose to surface winds. Also, while oceans tend to remain equable in temperature at all hours and all seasons, continents tend to get too warm in summer and too cold in winter. As cold areas tend to develop high surface pressure, and warm areas low surface pressure, the idealized latitudinal pressure belts of the northern hemisphere are in fact somewhat broken up into continental and oceanic 'highs' and 'lows' which change considerably with the seasons. These semipermanent, or seasonal, lows and highs are called *centers of action,* because their counter-clockwise or clockwise circulations control in some degree the general movements and conflicts of air masses in their neighborhoods, and influence greatly both the position and movement of various segments of the polar front and the development and motion of the traveling cyclones that are largely responsible for most temperatezone bad weather.The centers of action in and around North America — those that have most bearing on our weather in the United States -are as follows:

Winter	Summer
Canadian High	Hawaiian High
Aleutian low	Bermuda-Azores high
Icelandic low	(Rokies low - weak)
	(Aleutian and Icelandic lows - weak)

It will be well to recall these centers of action, and to remember the cardinal fact that (in the northern hemisphere) winds blow clockwise about a high-pressure area and counter-clockwise around a low-pressure area — particularly when we come to consider the traveling 'highs' that are usually associated with vagrant air masses, and the traveling 'lows' or cyclones out of which grow warm and cold fronts.

Table. Typical climate of the deep tropics

Month	Mean Tempera-ture	Mean Cloudiness (% of sky)	Monthly Rainfall (inches)	Average Wind direction and force
Jan.	80	50	3·5	North, moderate
Feb.	80	52	1·5	North, moderate
Mar.	81	50	1·5	North, moderate
Apr.	81	61	4·2	North, light
May	81	75	12.6	Variable, light
June	81	81	13·7	Variable, light
July	81	81	15.6	Variable, light
Aug.	80	80	15.2	Variable, light
Sept.	81	77	12.5	Variable, light
Oct.	80	77	15·5	Variable, light
Nov.	79	79	22.2	Variable, light.
Dec.	80	61	11.1	North, light

What are the outdoor conditions of life and thought and feeling near the poles and within the arctics? As a brief and partial answer to this always interesting question, Frederick Crockett, who was also on the 1928-30 Byrd expedition, gave me the following word-picture:

Usually during the months-long arctic 'day' there are low-hanging clouds, but below them, along the ground, visibility may be very good. The view in all directions is extraordinarily uniform — no shadows, no dark objects, no horizon, all gray above blending with chalky white below, 'like walking in a bowl of milk.' On this white, featureless surface it is impossible to see hummocks before stumbling over them, and walking is consequently very tiresome without skis, which more or less 'feel their way' along.

Distances in this milky light are very deceiving. A piece of dog dung a hundred yards away may look like a far-distant house. Because of the uniform white glare from the surface below, from which men's eyes are not naturally protected, sun blindness is likely unless protective goggles are worn. A very painful experience once it is incurred, something like burning hot sand in your eyes.

We had no trouble sleeping safely out on the trail at temperatures around 50° F. below zero. Our sleeping bags were caribou hide with the hair turned inward, with thin woollen blankets inside. When crawling out in the morning one had to exhaust the bag of its warm air, lest the vaporous moisture condense and freeze, making a very uncomfortable sleeping place for the next night.

For clothing, we departed somewhat from previous arctic experience and wore separate thin layers of wool (successive sweaters). These sweaters we put on or took off, according to temperature and the degree of exertion needed at the time. On the trail we carried an outer natural-fur parka, but did not usually wear it (even at fifty-below temperatures) until we made a long halt, or when outside in the evening, and so on.

In general, wind was more important than temperature as a factor in comfort. On one winter day in July, 1929, two other men and I started out on skis for a short local exploration. It was 9 A.M., and still completely dark, when we left camp. The temperature was colder than 70 below -actually -72° F., as I remember it. Fortunately, however, there was no wind. It was completely calm; the sky was clear, with the intensely bright southern-hemisphere stars (Canopus, the Southern Cross, and so on) swinging slowly around over our heads as the morning wore on, interspersed with the pale-green, flickering curtains of the aurora australis.

Starting off cool in thick clothing, we got well warmed up in a half-hour of skiing across the sand-like antarctic snow, which does not melt slightly under your skis like snow in warmer climates, providing a slippery surface. Hence skiing calls for heavy exertion in order to make any progress. Strange as it may seem, though at 9·30 A.M. the temperature was still 70 below and it was still dark, we were going forward barehanded and almost stripped to the waist. The only time we really felt the cold was on slight downslopes where we could slide along at six or seven miles an hour. Here the lack of heavy exercise, combined with the slight wind of our forward speed, made us immediately feel the bitterness of the Antarctic cold. We continued thus in the darkness until around noon, when the sun appeared for an hour or so as a faint glow on the northern horizon.

To find the day-to-day march of normal temperature in your own particular American region, you must consult a Weather Bureau publication entitled 'Normals of Daily Temperature of the United States.' But by averaging the normals for Washington City with those of certain inland cities in the same latitude, we can obtain something fairly representative of a large (central and eastern) portion of the country.

Atlantic seaboard temperatures average two or three degrees Fahrenheit warmer in winter than inland temperatures for the same latitude; in summer the seacoast may average a degree or two cooler. Also, yearly maximums and minimums on the seacoast come several days later than corresponding maximums and minimums inland.

The northern United States (latitude of Boston) averages about five degrees Fahrenheit colder than this table. The southern United States (latitude of Savannah) averages about ten to fifteen degrees warmer than this table in winter and about three to five degrees warmer in summer. Temperatures in the Rockies and other high mountain regions vary widely with altitude and topography. Temperatures on the Pacific coast are largely conditioned by the water temperatures of the eastern Pacific Ocean, and by topography. In general, Pacific seaboard temperatures show far less annual change than inland temperatures, and yearly maximums and minimums are delayed a month or so longer. Summer is typically cool and foggy, because of cold sea water that wells up from the great depths along the west coast.

The normal-temperature table printed above shows many well-known but interesting facts. The time of greatest cold comes around January 25, lagging more than a month behind the sun. The temperature remains almost constantly cold for a while; then, with the coming of spring, it begins to rise, slowly at first, later at the rate of S!° to ½° F. per day or 10° to 15° F. per month. But the month of April is nearly as cold as the month of November; though many people, rushing out scantily clothed into the brightening sun, fail to realise the fact, and return indoors to meditate on it while nursing a spring cold. In the northernmost parts of the United States winter lingers into April, perhaps into May. A period of nearly uniform summer heat comes in July and August, the maximum (around July 25) again lagging more than a month behind the sun, which has already turned back southward; and this summer heat is surprisingly uniform, by and large, over the whole United States. Then the temperatures begin to fall, slowly at first, then more rapidly, until by midautumn the rate of decrease is about equal to the rate of increase in spring. Finally the winter snows begin, and the snow itself, during the ensuing months, tends to hold air temperatures below freezing by reason of its reluctance to melt and its high reflecting power.

Tabulated observations at various levels show the following prevailing directions and velocities: altitude 1 mile, mostly westerly, averaging about 15 m.p.h. in summer and about 25 m.p.h. in winter; 2 miles, mostly westerly, averaging around 18 to 28 m.p.h.; 4 miles, mostly westerly, averaging about 25 to 40 m.p.h.; 6 miles, westerly to west-northwesterly, averaging about 40 m.p.h. More detailed information on these American upper prevailing winds is contained in a pamphlet published by the United States Weather Bureau.

Chapter 7

Storms of the Temperate Zone

COLD AND WARM AIR MASSES

We hear a great deal, in these machine-motivated days, about 'air-conditioning' — by which it is meant that a houseful of air, or a trainful or an airplaneful, can now be adjusted in both temperature and humidity to suit individual human tastes or general human needs. Some mechano-prophets, unmindful of Nature's wholesomest pleasures, even want to make all our houses windowless.

All this air-conditioning talk is nothing very new to Nature, who has herself been in the air-conditioning business for a good many thousands, or a good many millions, of years. But Nature seldom bothers with a houseful of air — she is more likely to do the job by the continentful or the oceanful, and her conditionedair masses are usually several miles deep and several hundred miles wide.

This open-sky air-conditioning is what we mean, in general, by the 'air-mass analysis' emphasis that swept over all the meteorological circles in the world during the nineteen-twenties and thirties. So far as temperate-zone weather is concerned, no more important viewpoint and method of analysis was ever propounded.

Air-Mass-and-Front Analysis

Air-mass-and-front analysis really began during the First World War of 1914-18. About 1903, it is true, Frank Hagar Bigelow, of the United States Weather Bureau, had developed his 'counter-current' theory of the origin of cyclones, amplifying the earlier ideas of Dove. Bigelow thought that the interchange of air between equator and pole takes the form of great currents of unlike air, a warmmoist tropical current flowing side by side with a cold-dry polar current.

In England, Admiral Fitzroy had in some degree sensed the cold and warm sectors of a cyclone by 1863; and Napier Shaw noted cyclonic wind-shift lines about 1905. In 1918 United States Army weather forecasters with the American Expeditionary Force in France used technique closely approaching modern physical weather analysis. But chief credit for the systematic building of all these early ideas into the unified and workable whole of modern air-massand-front theory and practice — a picture of 'fronts,' 'advances,' and 'retreats' that reflected the terminology of the First World War — must go to a few weather analysts in Norway.

Three of them are from one lineal line: Carl Anton Bjerknes, the grandfather, began with theoretical hydrodynamics in the nineteenth century; Vilhelm F. K. Bjerknes, the father, applied hydrodynamical theory to meteorological theory; Jakob Bjerknes, the son, translated this theory into practical applications. Collaborating with the younger Bjerknes (who, incidentally, came to America in 1939, and settled at the University of California) were others of the Bergen school — notably Halvor Solberg, Tor Bergeron, and Sverre Petterssen (who also moved to America in 1939). And it is appropriate, in a way, that these concepts of conflicting warm and cold air masses should have come to fruition in Scandinavia — for Norse mythology resounds with fierce and never-ending conflicts between Thor, God of the lightning and the beneficent summer heat; and the jotun Rime, demonaic giant of frost and cold.

The Bjerknes 'polar-front picture' first appeared in 1918. During the winter of 1923-24, Vilhelm Bjerknes, just beginning to get reliable routine results at Bergen, came to the United States and lectured at several institutions; this was the first thorough personal exposition of air-mass-and-front analysis to reach America. In 1926 Carl-Gustave Rossby, a Swede active in the Bergen school, also came to America and, failing to interest the United States Weather Bureau in the new theories at that time, went to Massachusetts Institute of Technology, where as professor of meteorology he made important contributions to the science.

The new ideas did not appeal at once to every conservative meteorologist accustomed to other viewpoints, and in some quarters they were tacitly or outspokenly opposed. But their present universal use leaves little doubt that they represent a truer and more complete system of viewing temperate-zone weather phenomena than has ever existed before.Air-mass-and-front analysis is based in general on five main ideas, simple enough in themselves, but wonderfully complex in their complete application. These basic ideas are:

- When air stagnates for some time over a distinctive geographic area, called a 'source region,' it acquires characteristics, such as temperature and moisture, in keeping with the current climate of the source region. For example:

'polar' air is relatively cold, 'tropical' air is relatively warm, 'continental' air is relatively dry, 'maritime' air is relatively moist. In general these characteristics are not limited to air near the surface, but extend several miles aloft.

- Impelled by the general atmospheric circulation, large masses of air (hundreds of miles across and several miles deep) may travel long distances over the earth's surface, still retaining in some degree their source-region characteristics, which are modified only gradually as the air mass sojourns in a new geographic region. This modification is usually accomplished first at the surface, later aloft.
- When two dissimilar air masses are brought together by the general atmospheric circulation they do not immediately mix; a definite boundary or discontinuity zone (perhaps five or more miles wide) occurs between them. Along and near the boundary colder (denser) air always underlies warmer (rarer) air in the form of a wedge; hence the 'front' between the warm and cold masses (actually a mixing zone but pictured and spoken of as a discontinuity surface) is usually inclined at a small angle to the earth's surface.
- Definite air-mass interactions (producing variable winds, clouds, rain, and other weather phenomena) occur along these fronts. The form and intensity of these interactions depend on the differences in characteristics of the dissimilar air masses involved, their extent, and the rapidity with which they are brought together.
- In addition to this 'frontal' weather, clouds, rain, and other weather effects may occur within air masses where the characteristics of the air mass itself favour them, and where terrain features and the motion of the air mass accentuate these characteristics.

It must be emphasized again that air-mass-and-front analysis applies mainly to the temperate zones of the earth's surface. The tropics, as we saw in the preceding, are under warm, moist, 'tropical maritime' air almost all the time, and this air varies in heat and moisture only slowly with the seasonal, north-south slow swing of the apparent sun.

The arctics are likewise under fairly uniform, cold, fairly dry 'polar continental' air most of the time. It is only in one of the temperate zones that a real 'polar front' marks the continuous interplay of unlike air masses and large-scale cyclonic storms. However, the temperate zones in general, and the north-temperate zone in particular, are where most of the world's people live, and where most of mankind's intellectual and material advancement is carried on. Most of our weather in the United States is controlled directly by air masses and fronts.

Air-mass-and-front analysis, unfortunately, is not intrinsically a simple subject. But these fundamental principles of temperate-zone weather, too new at present to be adequately treated by most literary aerologists, are far too important to be slid over with anything less than complete understanding even in a book addressed primarily to laymen.

Cold and Warm Air Masses

Any air-mass classification, anywhere, must take into account two general types of air with regard to temperature, and two general types with regard to moisture. There are relatively cold air masses and relatively warm air masses. There are also relatively dry air masses and relatively moist air masses. All this adds up to just four basic types of air masses — to be found, at one time or another, almost everywhere in temperate latitudes, as follows:

Cold-dry (Symbol $C_{D)}$ Warm-dry ($W_{D)}$

Cold-moist ($C_{M)}$Warm-moist ($W_{M)}$

This rudimentary air-mass classification is possibly sufficient for most laymen who want to talk intelligently about the weather. Much of the time, indeed, it may even serve for amateur aerologists who, in the United States at least, can easily draw weather maps of a small and rather sketchy sort from coastal or airways radiophone reports.

But there are certain advantages, particularly serious aerologists who do more responsible forecasting and deal with larger and completer weather charts, in an air-mass classification having geographical significance. In general form, good anywhere in temperate latitudes, this geographical classification is as follows:

Polar continental (Symbol (P C) — corresponds to cold-dry.

Polar maritime ($P_{M)}$ — corresponds to cold-moist.

Tropical continental or subsidence (T_{C}, T_{S},) — corresponds to warm-dry.

Tropical maritime ($T_{M)}$ — corresponds to warm-moist.

The two tables given above are the whole alpha and omega of air-mass analysis in one very qualitative nutshell. If you want to know anything at all about the great science of temperatezone weather, you must know these.

Before air-mass-and-front analysis could be fully and thoroughly applied to weather situations in the United States, it was first necessary to find out what the most frequent North American air masses were, and to study their characteristics. This fundamental research was accomplished at Massachusetts Institute of Technology by Professors Rossby and Hurd C. Willett; and its results were embodied in Willett paper on *'American Air-Mass Properties'* in 1933.

The Rossby-Willett air-mass classification is based on the various American source regions from which our prevalent air masses — polar and tropical, continental and marine — are known to come. American polar continental (cold-dry) air, for example, comes actually from a conditioning sojourn over northern Canada, Alaska, or the Arctic ice cap; therefore we can call it 'polar Canadian.' Much of our warmest and moistest (tropical maritime) air in the United States comes from over the Gulf of Mexico and the Caribbean Sea; hence we can call it ' Tropical Gulf.' Likewise with other air masses from other American source regions.

Our professional aerologists, who daily deal with weather maps covering all of North America as well as a multiplicity of diagrams representing upper-air conditions, must rigorously follow the movements of various air masses from day to day as they hurry or idle across this hemisphere. Hence the RossbyWillett air-mass classification, based on definite American source regions rather than climatic generalities, is a necessary tool in their work. It is also essential knowledge for any weather amateur who tries to draw or follow large weather maps and their daily changes.

With all these ideas in mind, we are now ready for one single table that shows at a glance the whole interrelationship of North American air masses, subject to some further description and elaboration.

The ebb and flow of these air masses across the United States and its surrounding seas is controlled largely by the high-low circulations of the North American 'centers of action' — the Aleutian and Icelandic lows, the Hawaiian and Bermudian highs. Most of our bad weather - cloudiness, high winds, rain, or snow — arises not so much from the air masses themselves as from the incessant conflicts between unlike air masses inexorably converging upon one another. Air-mass conflict results in the formation of cyclones and fronts — the large-scatt storms of the temperate zone.

Polar Pacific (P_{P}) air masses originate over the Bering Sea and the northern Pacific Ocean. They are characterized in summer by moderate temperature and humidity, and in winter by moderately low temperature and moderate humidity. They are usually rather unstable, becoming very unstable in winter and spring, when their progress across the western United States is attended by heavy cumulus clouds and precipitation. East of the Rockies, modified Polar Pacific air (N_{PPC}) is much drier and somewhat warmer, particularly at lower levels.

Over the central United States in summer, Polar Pacific air is much like polar continental air. Polar Pacific classifies, in general, as polar maritime (P_{M}) or cold-moist (C_{M}) air.

Polar Atlantic (P_{A}) air masses originate over the colder portions of the northern Atlantic Ocean. They are characterized by low to moderate temperature and rather high humidity, some instability at lower levels, and low stratus clouds. They seldom extend far south or west of New England; and even here they are rather rare, because their invasions depend on an east-west (or reverse) wind gradient in temperate latitudes. Their general type is polar maritime (P_{M}) or cold-moist (C_{M}).

Tropical continental (T_{C}) air masses, which appear only in summer and then rather rarely, originate over Mexico and the Southwest. They are characterized by very high temperature, moderate to low humidity, and generally clear skies despite considerable instability. But warm, dry air masses having T_{C} characteristics often appear aloft over wide regions of the United States in winter and summer alike. These upper-air masses originate at high levels in the subtropical belt of high pressure — that is, in the horse latitudes around lat. 30° N.

Tropical subsidence or 'subsiding' (T_{S}) air masses are thus of far greater importance in North America than the rarer surface-conditioned T_{C} air that they resemble in characteristics and weather. type. The T_{S} air is upper-air-conditioned by reason of the fact that, in any large high-pressure area, the dome-shaped accumulation of air (represented by high or increasing surface pressure) is gradually but continually sinking vertically and spreading out laterally.

The North American seasonal highs or centers of action — the Hawaiian high and the Bermuda-Azores high — are no exceptions to this general subsidence rule. Hence the slowly sinking air in them is gradually warmed (and gradually dried, at least in relative humidity) by its very subsidence. Sometimes, as in the summer 'heat waves' of the Midwest, T_{S} air over the United States extends all the way down to the surface. Common North American T_{S} air and the rarer T_{C} both classify generally as warm-dry (W_{D}).

Tropical Gulf (T_{G}) air masses originate over the Gulf of Mexico and the Caribbean Sea. Tropical Atlantic (T_{A}) air masses originate over the Sargasso Sea portion of the Atlantic, a weed-littered largecurrent eddy north and east of the West Indies. Both types of air masses have similar characteristics. They are characterized in summer by high temperature and very high humidity and in winter by fairly high temperature and high humidity.

They are conditionally or convectively unstable at all levels, easily forming heavy cumulus clouds with showers and thunderstorms in summer, and extensive cloud-and-precipitation areas in winter. T_G and T_A air is, in fact, responsible for most of the warm, 'muggy,' thundery summertime weather of the southern and eastern United States. In winter these air masses may bring general fogs at the surface; and when widely lifted by frontal action, they invariably form widespread areas of heavy clouds, rain, and/or snow. Both T_G and T_A classify generally as tropical maritime ($T_{M)}$ or warm-moist ($W_{M)}$.

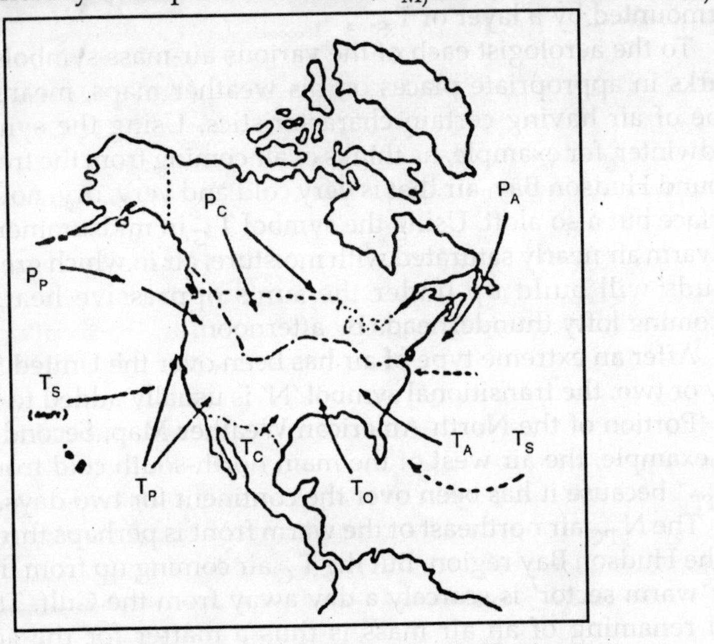

Fig. 44. North American Air-Mass Source Regions

Tropical Pacific ($T_{P)}$ air masses originate over the North Pacific Ocean trade-wind belt northeast of Hawaii. They are less warm, less moist, and more stable than T_G or T_A air; and are characterized by moderate temperature, fairly high humidity, and neutral stability, forming stratus clouds if lifted moderately, and heavy cloudand-rain areas if forced to rise far enough by mountains or fronts. T_P air is found mostly along the Pacific Coast, but sometimes extends east of the Rockies at high altitudes. It causes most of the frequent heavy rains of the Pacific Northwest, and sometimes appears at the surface along the California coast. Classified generally, it is a moderate form of tropical maritime ($T_{M)}$ or warm-moist ($W_{M)}$.

It must never be forgotten that all air-mass-and-front analysis is three-dimensional. It is a matter not merely of surface weather observations, but also of simultaneously considered upper-air observations, direct and indirect — airplane or balloon soundings for temperature and humidity, and pilot-balloon observations of upper winds, where these are possible; where not, at least careful eye observations of clouds at all visible levels (their type, appearance, development, and motion) and of any precipitation that falls from them. In this three-dimensional picture, more than one air mass may be present over a single station. There may, for example, be a shallow layer of P_c, overlaid by another layer of T_G, which is in turn surmounted by a layer of T_S.

To the aerologist each of the various air-mass symbols, which he marks in appropriate places on his weather maps, means a distinct type of air having certain characteristics. Using the symbol P_C in midwinter, for example, he thinks of air coming from the frozen wastes around Hudson Bay, air that is very cold and very, dry, not only at the surface but also aloft. Using the symbol T_G in midsummer, he thinks of warm air nearly saturated with moisture, air in which great cumulus clouds will build up under the sun's oppressive heat, probably becoming lofty thunderheads by afternoon.

After an extreme type of air has been over the United States for a day or two, the transitional symbol 'N' is usually added to its symbol. On 'Portion of the North American Weather Map, Second Morning', for example, the air west of the main north-south cold front is called 'N_{PP}' because it has been over the continent for two days.

The N_{PC} air northeast of the warm front is perhaps three days out of the Hudson Bay region. But the T_G air coming up from the south in the 'warm sector' is scarcely a day away from the Gulf. The naming and renaming of an air mass is thus a matter for the aerologist's judgment. If he himself knows what he means by a symbol in the way of air-mass characteristics and weather, fine points of nomenclature are unimportant; particularly so, because complete air-mass analysis includes much more than the qualitative name.

AIR MASSES MOVE AND CHANGE

Conservative Air-Mass Qualities

Measure of atmospheric moisture which remains constant however the air in question may rise or sink, provided no condensation occurs. This measure is *specific humidity* — the actual mass of water vapour, in grams, present in a kilogram of air. Ordinary values of specific humidity in the open air may be anywhere between 1 g/kg and 20 g/kg. Since specific humidity in any air sample or layer is unchanged by rising or sinking, heating or cooling (provided no water condenses or evaporates), it is a 'conservative' measure of moisture. If air masses are to be identified, analyzed, tagged, and followed in their free motion about the air ocean — in their comings and goings, risings and sinkings — a conservative measure of temperature is likewise needed. This measure is *potential temperature*, and it is derived directly from the dry adiabatic cooling rate (10° C/km or about 5° F/1000 ft) of rising air.

Suppose that the temperature of an air sample at a height of kilometre (3300 feet) is just freezing. This temperature may be expressed as 32° F., 0°C., or 273° Absolute. It may also be expressed as potential temperature on the Absolute scale — in this case, about 283° A. This potential temperature simply means that, if the air sample in question is compressed (or rarefied) until it is at standard sea-level pressure, its temperature will be 283° A. Hence the definition of potential temperature — the temperature that an air sample would have if its pressure were changed to sea-level standard.

If an air sample is well up in the free air, its potential temperature is of course higher than its actual temperature — because to reach sea-level pressure it must sink, be compressed, and be heated. If the sample is at the bottom of a deep mine, however — whence it must rise to reach sea level — its potential temperature is necessarily lower than its actual temperature.

Once an air sample is tagged at a certain potential temperature, this quality will not change howsoever the air sample rises or falls, expands or contracts. Provided, that is, the air sample receives no heat from external things (which possibility can usually be disregarded, and usually has to be, in the interests of simplicity). And provided also that no condensation or evaporation of moisture occurs.

Air-Mass Identification

It is now evident that, with conservative measures of both temperature and moisture, not only air samples but air layers, and even large air masses, can be tagged, analyzed, and to some extent kept track of in their wanderings if repeated soundings are made into them. As a result of these soundings, potential temperature and specific humidity can be determined at all levels, and the results can be plotted on a potential temperature — specific humidity (è — *q*) diagram. This diagram, devised by Rossby at Massachusetts Tech. in 1931, has potential temperature as a vertical ordinate and specific humidity as a horizontal ordinate.

Vertical lines on the *è-q* diagram are labeled with values of specific humidity such as 5 gm/kg, 10, and so on. Horizontal lines of constant potential temperature are numbered 240° A., 270, and so on. Other lines, which need not be considered here, are also printed on the diagram. Of course, constant potential temperature through an air layer means the same thing as a dry adiabatic (about — 5° F/1000 ft) vertical temperature gradient or lapse rate measured in ordinary temperatures. Hence the slope of the sounding curve, as plotted on the *è-q* diagram, tells much about the stability of the air layer that was sounded through. In particular, the diagram shows at a glance whether or not the air in question is convectively unstable.

But perhaps the main use of the *è-q* diagram is in identifying different air masses as they drift across the weather scene. Colddry air (of polar continental origin) is marked by low and fairly constant values of specific humidity, together with a large increase of potential temperature with height — all of which means that the curve of a sounding made through this type of air will fall in a certain and typical place on the *è-q* diagram.

Warm-moist air (of tropical maritime origin) is marked by fairly constant potential temperature, and by very high specific humidity at low levels, decreasing rapidly upward. Hence its sounding curve will appear in a typical and entirely different place on the *è-q* diagram. Thus when soundings made over a wide area are plotted on the *è-q* diagram, the air masses within this area can be identified, and their driftings can be followed from day to day.

A competent American aerologist and forecaster, in fact, may well have in front of him a model *è-q* diagram appropriate to the season (winter, spring, summer, fall) on which typical sounding curves of all the usual North American air masses -polar Canadian, polar Pacific, polar Atlantic, tropical continental, tropical subsidence, tropical Gulf, tropical Atlantic, and tropical Pacific — are plotted. By means of a translucence frame, the sounding of any new and questionable air mass can be quickly compared with the typical curves (making due allowance for seasonal and other air-mass modifications); and the new air mass can thus be identified and labeled on the map.

Air-Mass Modification

It is inevitable that extreme types of air, moving out of their tropical or arctic source regions into temperate latitudes, will be gradually modified by the changing climatic conditions that they encounter. What is the sort and extent of this modification? Whatever it may be, by all the laws of aerology any modification resulting from a differently warmed or cooled or watered or dried surface must first appear in the lower levels of a wandering air mass, then gradually extend farther and farther aloft with the development of vertical turbulence or up-and-down air currents. Cold polar air, moving southward, usually finds the land or water surface getting gradually or suddenly warmer beneath it. Hence the most common modification of southward-moving polar air is a pronounced heating of the lower levels. This invariably steepens the lapse rate, and makes the air mass as a whole more and more unstable (or at least less and less stable) as it travels southward.

If the southbound polar air travels over a surface that is not only relatively warm, but also wet — a water surface, in other words — this increasing temperature-instability is usually further complicated by increasing moisture-instability. The immediate result is cumulus clouds, showers, and generally squally weather, which will be still further aggravated if mountains lie beyond the water. Thus a violent outbreak of cold P_c air, rushing southeastward in wintertime out of Canada across the Great Lakes, invariably produces heavy clouds along the lee lake shores, and heavy snow squalls on the windward slopes of the Alleghenies. The same outbreak may later produce clouds and showers out over the Atlantic Ocean, and particularly over the Gulf Stream.

When cold, moist air advances over a mountainous land sur face, it loses moisture in the form of rainfall: and, if the land is warmer than its ocean source, the air gradually gets warmer and drier. Thus springtime eastward outbreaks of cold, moist, unstable, P_P air lose much of their moisture in the form of showers along our mountainous Pacific coast, and continue eastward, over the Rockies, as a much drier N_{PPC} additionally warmed by the foehn-wind effect of a long glide down the leeward (east) slopes of the continental backbone.

Warm tropical air, traveling northward over a relatively cold surface, is of course progressively cooled in its lower and middle levels. Hence its lapse rate is decreased, and it becomes more and more stable, and less and less inclined towards vertical currents and general storminess. If the tropical air is very dry (T_C or T_S), it remains clear; and the end result (except in midsummer) is fine, fair, mild weather — a delight to all outdoorminded creatures. In midsummer, the not-so-pleasant result may be a heat wave.

But if the northbound tropical air is very moist (T_M of some sort — say T_G or T_A), its lower levels are soon cooled below their dewpoint, and the result (in winter, at least) is widespread fog of the advection type. The height of this fog depends on the surface and lower winds — the speed with which the whole warm-moist air mass is drifting. If the winds are very light, there is little turbulent mixing in the lower levels, and the fog is perhaps less than a hundred feet thick.

If the winds are stronger — up to fifteen or twenty miles per hour — the fog layer may extend through several hundred feet, and may be lifted several hundred feet from the surface as 'high fog' or stratus cloud, also, if the northwardmoving fog encounters a warmer surface, it may lift off the ground into low stratus. The most common examples of these warm-moist air-mass fogs are the notorious and frequent Grand Banks fogs off Newfoundland, where the water surface is unusually cold, and the occasional widespread wintertime fogs of our Atlantic littoral, formed when tropical air blows in from the ocean over the snow-covered land.

Aside from the modifying effects of the surface beneath them, air masses can be considerably changed by lifting or sinking. In any cold-air-mass centre (any high or increasing pressure area, that is) — or at least in the slower moving ones — sinking or subsidence of air is continually going on. The average subsidence rate is something like half a mile per day. No great amount of computation is necessary to show that the upper levels of any such typical cold-air mass should get warmer by ten or fifteen degrees Fahrenheit every twenty-four hours.

The lower levels nearer the ground do not share in this warming, being cooled by ground radiation at night and, more important, being incapable of sinking much anyway. As a result, stagnant polar continental air, and the typical slow-moving high-pressure areas associated with it, show marked temperature inversions ('subsidence inversions') that are most noticeable during the late night and early morning hours.

Air masses can also be modified a good deal by lifting -usually inclined-plane lifting on a land-surface slope or a wedge of colder underlying air. The most usual net effect of such lifting — at least in a warm-moist air mass — is a steepening of the existing lapse rate. Hence the lifted air becomes less stable, and if it was convectively unstable to begin with, it will soon become actively unstable with the formation of clouds and rain or snow.

The underlying reason for this steepening of the lapse rate with lifting is that the lower part of air layer, being usually moister than the upper part, begins condensing into cloud while the upper part is still clear. Hence, with further lifting, the lower part cools merely at the moist-adiabatic (about 21/2° F/1000 ft) rate, while the unsaturated upper part cools at the dry-adiabatic (about 5° F/1000 ft) rate.

Anticyclones

It is apparent from United States weather maps that a mass of cold polar air is usually associated with an 'anticyclone' or high-pressure area. At first glance the reason would seem to be that, as cold air is heavier than warm, the greater weight of an overlying mass of cold air causes surface barometers to register a 'high.' The actual picture, however, is more complicated; nor can we examine, here, any but its more general aspects.

A mass of cold air usually takes the form of a wide, shallow dome — somewhat the way a drop of water on an oily table spreads out into a lens-shaped disk. The cold dome is usually three or four miles high (or rarely, as much as seven miles high, reaching clear up to the stratosphere base), and several hundred miles wide. Above the cold dome, usually, is relatively warmer (though actually cold, of course, at these heights) air of tropical origin.

The top of the cold dome does not coincide at all, strangely enough, with the centre of high pressure at the surface — because the *surface high represents*, not the greatest amount of cold air aloft, but *the greatest convergence or crowding of air aloft*. Usually the dome cap is several hundred miles ahead (east) of the 'high' centre. This discrepancy has led many aerologists to contend that surface highs are 'made,' so to speak, in the stratosphere -by a convergence or accumulation of more than ordinarily cold air at these great heights. It is quite certain, anyhow, that any surface high which builds up suddenly on the weather map, without visible aid from air-mass transport in the lower levels, must originate from the convergence of air in the upper levels.

Stratospheric winds, beyond reach of ordinary pilot-balloon runs, often blow at one hundred to two hundred miles per hour. Yet the total mass of the atmosphere above ten miles is only one ninth of the mass below this level, a fact that throws some doubt on this theory of 'upstairs' control. The whole question will not be completely solved until both our airplane or radioballoon soundings and our pilot (or ray-pilot) balloon upperwind runs, as a matter of daily routine, extend considerably higher than heretofore.

But getting back to the structure of the cold dome: the *dome cap doer coincide* more or less on the weather map *with a sort of dynamic centre of surface pressure* — the region of most active pressure rise on the front edge of the high. The high itself is outlined on the map in isobars — lines of equal surface (reduced to sea-level) pressure. The 'dynamic high,' more directly under the cold dome and perhaps more directly caused by it, is outlined on a weather map by isallobars — lines of equal pressure change.

One factor that can never be forgotten in any of highs and lows, air masses and fronts, is *air-mass momentum*. Even a single cubic mile of 'thin air' just above the earth's surface weighs something like five million tons! A wintertime out break of polar air, say, may be several hundred miles wide, and may comprise many billion tons of cold air, all moving southward at perhaps thirty to fifty miles an hour. The amount of kinetic energy, or momentum, involved in the air-mass motion is evidently enormous. The net effect of this momentum will be to keep air-mass movements, once started, going until they are slowed down or stopped by friction – a process that usually takes several days.

BLOW HOT, BLOW COLD

Among all the weather vagaries that soothe or irritate Earth's creatures, few can be more irritating, depressing, or stimulating than violent extremes of temperature, particularly if inordinate heat or cold follows untimely from the opposite extreme. These sudden temperature changes, which haunt temperate latitudes (Particularly Lats. 40° to 50°) in maximum frequency and ferocity, are called 'heat waves' and 'cold waves.'

Alaska, lying between 60° and 70° north latitude, may seem a queer place for a heat wave. Nevertheless they do happen, relatively speaking, in our great northwestern territory. In early December, 1934, an unprecedented northward invasion of warmmoist (tropical Pacific) air caused heavy rainfall along the south Alaskan coast and skyrocketed inland temperatures from their usual minus winter readings up into the plus-fifties. Fairbanks was around 80° F. warmer than normal, with +58° F., as balmy air currents invaded the arctic twilight zone.

In the continental United States, and most particularly in its Midwestern section, a prolonged heat wave may be serious indeed. Warm-moist (tropical Gulf) invasions bring extremely 'muggy' weather that often causes deaths, but the conditional instability of W_M air at least usually provides afternoon showers along with the heat. Far worse are the northward and northeastward flows of summer warm-dry air (tropical continental or tropical subsidence). When that great summer centre of action, the Bermuda-Azores high, is well developed and somewhat west of its usual position, it pumps sizzling W_D air masses up over the heart of the continent, under clear, glaring skies, for days on end.

The worst W_D air heat wave in recent United States history occurred in the summer of 1934, as part of a paralyzing drought that was indeed, world wide. From mid-June to mid-August, not a single day passed without bringing a temperature of 100° F. or higher somewhere in the Midwest. These torrid century marks were not at all limited to perennial desert hotboxes like Yuma, but spread over the whole country. In Oklahoma, Kansas, and Missouri, day after day, afternoon temperatures ran well over 100° F. And at night, while millions of Midwesterners tossed on sleepless beds, thermometers remained well above 80°. This great 1934 drought, to be blamed on persistent invasions of W_D air, took a terrible toll of plant life, and set the scene for scouring winds in later months and years — winds that would make a 'black hell,' from which thousands of homesteaders were to flee in despair, of the great American 'Dust Bowl.' Millions of trees also died in what amounted to a slow and flameless, yet scaring, continent-wide forest fire, and the number of real forest fires was vastly increased. Many ponds and lakes dried up entirely.

The Colorado River fell to its lowest stage, and the Red River of the North was completely dry in places where steamboat keels had earlier cleared bottom. The water level in parts of the Great Lakes, even, fell by a foot or two, impeding dredged-harbor navigation. Glaciers receded phenomenally on northern mountains. During the month of August alone, no less than fifteen hundred Americans died from the heat and causes initiated or aggravated by the heat.

Scarcely more pleasant than the summer heat wave is the winter cold wave. The Weather Bureau definition of a 'cold wave' varies somewhat for different parts of the country, but in general the term means a temperature drop of at least 30° F. within twenty-four hours or less, to a low around zero. New England is justly famous, or infamous, for the extreme variability of its weather (the usual cyclone tracks of most United States cyclones converge into one national exit right across these northeastern states). New England cold waves have been exhaustively studied by Professor C. F. Brooks and Mr. I. I. Schell of Harvard University. On New Year's Eve of 1933, for example, the temperature tumbled 55° F. as cold-dry (Polar Canadian) air rushed down out of the north on fifty-five-mile gales. On a February day in the same year, a P_c outbreak brought fifty-mile winds out of the north and northeast that sent thermometers tumbling by fifty degrees to below-zero minimum. On February 14, 1934, a seventy-mile blizzard drove the mercury down from freezing to six below zero.

Brooks and Schell found that clear warning of such New England cold waves could be read on the weather map, a full day in advance, by noting the temperatures and pressure gradients in Ontario and western Quebec. The colder the Canadian stations were, and the steeper the pressure gradient available to drive the northerly winds, the more violent the cold wave in store for New England.

Always more violent and changeable than maritime or coastal weather, however, are the rapid changes that take place in the middle of a large continent. Placed squarely athwart the P_c outbreaks just east of the Rockies is the elevated, tumbled tableland of the Black Hills, which have more weather in a week, at most seasons, than most stations have in a month. A slight taste of these Black Hills weather vagaries, as experienced on our stratosphere expeditions. Likewise in the matter of cold waves, Rapid City yields to few other United States stations. On January 5, 1930, the temperature was 56° F. at noon, zero at midnight, and continued down to — 9° F. by morning, the wind being only thirty miles or so out of the northeast. It was on the weather-historic date of January 12, 1911, that P_c air exercised its power on the town of Rapid City and its environs. An almost summery night, the mercury read +49° F. At 8 A.M., *two hour later, it bad tumbled sixtytwo degrees* to a low of 13°. That is probably the all-time coldwave record for the United States. A hunter who had left camp before dawn in ordinary clothing might easily have frozen to death before he could find shelter from the increasing northerly gales.

LARGE-SCALE STORMS OF THE TEMPERATE ZONE

Large masses of air, having distinctly different properties, may move long distances over the earth's surface without greatly changing their characteristics. The general earth circulation, driven by the stupendous long-range power of the sun and evidenced by trade winds, prevailing westerlies, semi-permanent centers of high and low pressure, and the like, favors the *convergence* of dissimilar air masses in certain earth regions, such as the northern Pacific Ocean and the northern and eastern United States.

Most active *cyclones* and *fronts* are formed in these convergence regions; but as the convergence regions themselves move north and south with the seasons, and fronts formed in them may remain active for days as they drift elsewhere, no part of the United States is free from occasional frontal activity of some sort. In some regions, such as Southern California and Florida, the general circulation favors the *divergence* of dissimilar air masses; in these places marked fronts are infrequent. As a result, also, of this general planetary circulation, most weather in northtemperate latitudes has an eastward drift (in the prevailing westerlies) averaging between fifteen and thirty miles per hour. (For rough weather-guessing, one can use 15 m.p.h. in summer, 2-5 m.p.h. at other seasons). If the weather always drifted eastward at constant speed in unchanged form, forcasting from telegraphic or radio weather reports would be child's play. But unfortunately the speed, the direction of motion, and the weather phenomena themselves are all constantly changing.

Development of a Temperate Cyclone

A temperate cyclone, or low-pressure area, does not 'just happen.' Usually it arises as a great atmospheric vortex between two still greater opposing air currents or air masses. Many temperate cyclones, perhaps most of them, first start as a great, long, horizontal wave on the surface of discontinuity, or *front*, between two unlike air masses that are moving unequally. We now trace the development of such a cyclone, bearing in mind that it is a rather typical case, here simplified, and often subject to many and large variations in the actual air ocean.

Suppose that in a convergence region two dissimilar air masses — cold-dry (polar continental) air on the north, and warmmoist. (tropical maritime) air on the south — are coming together. Along the front between the two, the cold air underlies the warm air in the form of a thin wedge, and the boundary surface slopes upward and northward. The cold mass moves eastward slowly, or perhaps has even a slight westward drift; the warm mass is hurrying eastward.

Along the front between the two air currents a vortex tends to form, in the same way that whirlpools are formed on a much smaller scale between adjacent currents of water moving at different speeds. The first indication of its formation is a wave in the formerly straight front. The warm current is bulging north into the cold current at the centre of the diagram, while the cold current is in turn bulging south into the warm current at both sides.

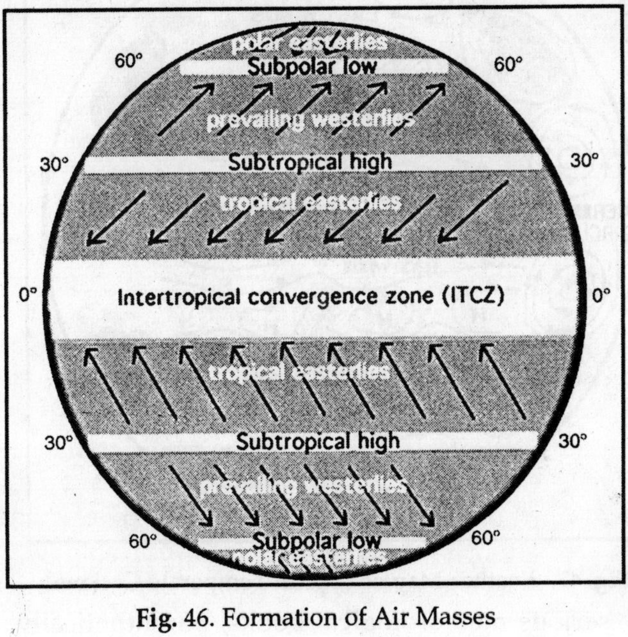

Fig. 46. Formation of Air Masses

To quote from the now classical 1922. paper by J. Bjerknes and H. Solberg: '*The essential condition for the formation of a cyclone is the coexistence of cold and warm air adjacent to each other*. The tendency of the cold air to spread along the ground, lifting the warm air, will then, with the modification caused by the rotation of the earth, create a cyclone and render energy to its growth.'

The next step in the formation of this vortex is a typical 'Bjerknes' cyclone. The tongue of warm air has extended well up into the cold current, forming a *warm* sector. East of the warm sector there is a *warm front*, beyond which the warm air continues to rush northeastward, climbing the slowly retreating wedge of cold air. West of the warm sector is a *cold front*, where a wedge of cold air, sweeping down from the northwest, is underrunning the warm air and encroaching on the warm sector.

The northward rush of warm air on the cast and the southward rush of cold air on the west produce between them a vortical *area of low pressure*, or *low*, which centers at the northern tip of the warm sector. It is indicated on a weather map by isobars, or lines of equal pressure, in the same way that a nearcircular valley is indicated by contours on a topographic map.

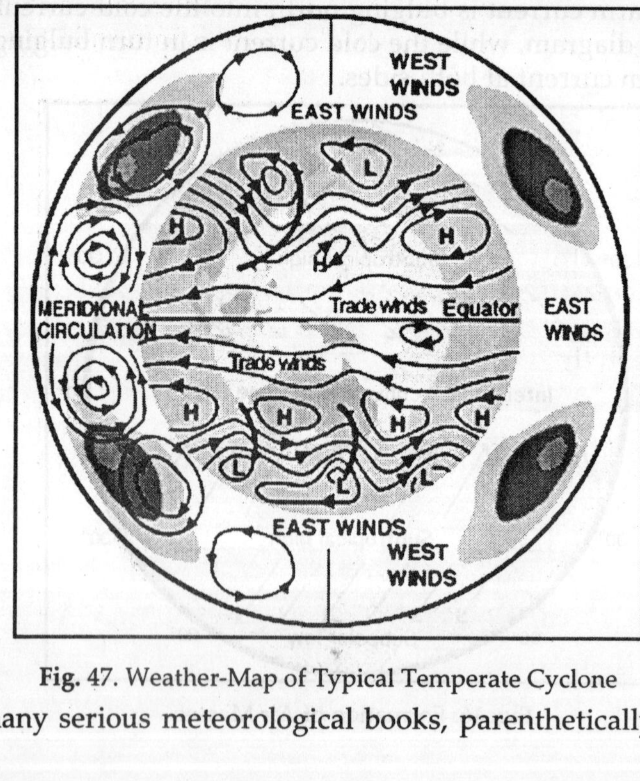

Fig. 47. Weather-Map of Typical Temperate Cyclone

In many serious meteorological books, parenthetically, a *low* is called an 'extra-tropical cyclone," in order to distinguish it from the rip-roaring tropical hurricanes. But this 'extra-tropical' terminology has always sounded clumsy to me, and in this book we shall continue to call these great temperate-zone storms simply 'temperate cyclones,' meaning not that they are necessarily mild, but that they haunt the temperate zone. Here might be mentioned, also, that the name 'cyclone,' applied to those terrible local air whirls known correctly as tornadoes is woefully erroneous though popularly common.

In the northern hemisphere winds blow about the low generally counter-clockwise: In the warm sector they are southwest or south. East and north (ahead) of the warm front they are south or southeast or east. West or northwest of (behind) the cold front they are west or northwest or north.

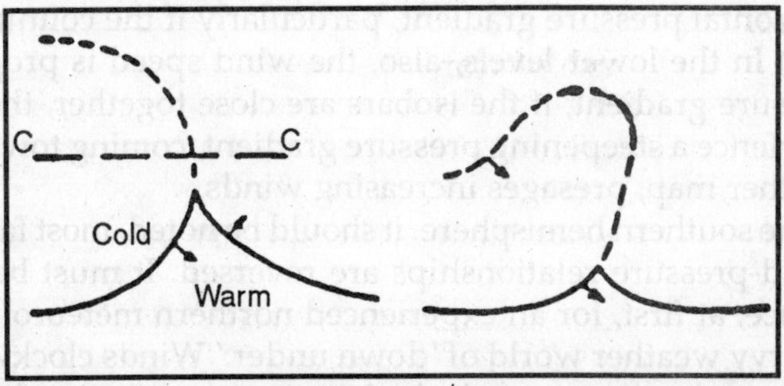

Fig. 48. Cloud-and-Weather Chart of a Typical Temperate Cyclone

The great, thousand-mile-wide cloud system around a large temperate cyclone is, properly viewed, a monstrous and marvelous thing — ahead of the warm front, cirrus thickening to cirrostratus, lowering and thickening to alto-stratus, then to high nimbo-stratus and lower scud (with precipitation), finally to heavy, lowering nimbo-stratus; in the warm sector, scattered fracto-clouds to clear or perhaps widespread fog or low stratus if the underlying water or land surface is cold; behind the cold front, a usually narrow zone of cumulonimbus, nimbo-stratus, and cumulus (with showers and squalls), thinning out to scattered cumulus to westward; north of the cyclone centre, mostly nimbo-stratus, stratus, and strato-cumulus rolls.

Behind the wedge-shaped cold front, the cold air mass has the shape of a shallow dome, perhaps a few miles high and several hundred miles across. Under the centre of the cold dome, where the piling up of cold air is greatest, there is the dynamic centre of greatest pressure increase, and a few hundred miles behind this there is the static centre of high pressure — the *anti-cyclone* or *high* — indicated on a weather map by isobars that resemble the topographic contours of a hill. If the high is moving rapidly southeastward behind a cold front (as many of them are), its winds, at least in the forward zone, are mostly northwest, and represent simply the southward or southeastward push of one large mass of cold air.

If the high is more or less stationary, however, the winds blow generally clockwise about its centre (in the northern hemisphere). In any case, an area of high pressure on the weather map represents relatively cool air piling up overhead to considerable depth in its forward portion.

The typical wind circulation about a well-developed low or a well-developed stationary high is often useful in predicting lower-level winds. Several hundred feet above the ground, these circulatory winds blow nearly parallel to the isobars. (Near the ground they veer toward the horizontal pressure gradient, particularly if the country is broken or hilly.) In the lower levels, also, the wind speed is proportional to the pressure gradient; if the isobars are close together, the winds are strong. Hence a steepening pressure gradient, coming towards you on the weather map, presages increasing winds.

In the southern hemisphere, it should be noted, most familiar high-low wind-pressure relationships are reversed. It must be a startling experience, at first, for an experienced northern meteorologist in this topsy-turvy weather world of 'down under.' Winds clockwise around a cyclone or low; counter-clockwise around an anticyclone or high. Warm, tropical air masses drifting up out of the north; cold waves rushing out of the south.

A warm front extending northeastward from its cyclone centre and advancing slowly southeastward, with the usual easterly winds ahead of it, then balmy northwest breezes with pleasant weather in the warm sector, then the violent cold front, extending northwestward from the cyclone centre and bearing down with showers and thunderstorms out of the southwest.

Before the advent of air-mass-and-front analysis, the cyclone perhaps a few miles high and several hundred miles across. Under the centre of the cold dome, where the piling up of cold air is greatest, there is the dynamic centre of greatest pressure increase, and a few hundred miles behind this there is the static centre of high pressure — the *anti-cyclone* or *high* — indicated on a weather map by isobars that resemble the topographic contours of a hill. If the high is moving rapidly southeastward behind a cold front (as many of them are), its winds, at least in the forward zone, are mostly northwest, and represent simply the southward or southeastward push of one large mass of cold air.

If the high is more or less stationary, however, the winds blow generally clockwise about its centre (in the northern hemisphere). In any case, an area of high pressure on the weather map represents relatively cool air piling up overhead to considerable depth in its forward portion.

The typical wind circulation about a well-developed low or a well-developed stationary high is often useful in predicting lower-level winds. Several hundred feet above the ground, these circulatory winds blow nearly parallel to the isobars. (Near the ground they veer toward the horizontal pressure gradient, particularly if the country is broken or hilly.)

In the lower levels, also, the wind speed is proportional to the pressure gradient; if the isobars are close together, the winds are strong. Hence a steepening pressure gradient, coming towards you on the weather map, presages increasing winds.

In the southern hemisphere, it should be noted, most familiar high-low wind-pressure relationships are reversed. It must be a startling experience, at first, for an experienced northern meteorologist in this topsy-turvy weather world of 'down under.' Winds clockwise around a cyclone or low; counter-clockwise around an anticyclone or high. Warm, tropical air masses drifting up out of the north; cold waves rushing out of the south. A warm front extending northeastward from its cyclone centre and advancing slowly southeastward, with the usual easterly winds ahead of it, then balmy northwest breezes with pleasant weather in the warm sector, then the violent cold front, extending northwestward from the cyclone centre and bearing down with showers and thunderstorms out of the southwest.

Before the advent of air-mass-and-front analysis, the cyclone or low used to be regarded as the underlying cause of its bad weather, and great attention was paid to the average seasonal tracks followed by lows of various types as well as their average rate of travel. Now it is seen that the low is itself only a result of greater underlying causes — a vortex formed along the front between unlike air masses by the convergence of their continentwide air flow. The driving and controlling forces that form the lows, and cause them to move in sometimes regular, sometimes erratic fashion, are to be found in the air masses themselves, in the anti-cyclones or highs associated with their centers, and in the 'centers of action' or great semi-permanent highs and lows that generally dominate hemispherical weather.

As to the motion of a low along the general cold-warm or 'polar' front, it can be resolved into two components. One is the *wave component* (always directed, in the northern hemisphere. about ninety degrees clockwise from the cold air, and hence usually eastward or northeastward), which is greatest for a wide, shallow warm-sector wave in the early stages of cyclone development, and may rarely give apparent speeds as high as one hundred miles per hour.

But as the cylone gets longer (from south to north) and narrower (from west to east), the wave component of its motion becomes negligible. There is also, however, the more tangible *wind component* of a cyclone's motion -simply the general or average mean of all the winds (at gradient and middle levels) blowing into and around the cyclone. In practice this wind component may be anything from a slight retrograde (westward) drift to thirty miles per hour or more eastward.

Failure to consider and evaluate these two components of cyclone motion has resulted in some notable weather-forecasting blunders. The day before President Taft's inauguration on March 4, 1909, forecasters in the Weather Bureau had their eyes on a low coming across country. It appeared certain that this cyclone would quickly pass the Capital, so they blithely issued an inauguration forecast, promptly relayed to the hundreds of thousands of visitors pouring into Washington, for bright and clear weather on the morrow.

Actually, the low retrograded after passing Washington, and brought back with it some of the nastiest weather the city had ever suffered. That very afternoon it clouded over. By nightfall it was raining, and by midnight this rain was freezing into glare ice. Next morning tons of white snowflakes fell over Washington. At the last minute Mr. Taft made a long speech in the Capitol Building, but while he was speaking the sun broke through with diamondlike intensity, and melted several inches of snow into a beautiful slush for the inaugural parade.

To everyone of inquiring mind who sees a large-scale storm, such as an intense temperate cyclone or a smaller but even more intense tropical hurricane, there must occur the question: Whence comes the enormous amount of energy evidenced in the winds and weather of such a storm? The answer is that the visible kinetic energy of cyclonic winds arises out of the potential energy of air-mass stability and distribution. In a tropical hurricane the air involved is very warm, very moist, and conditionally unstable, and continues to rise and draw inward because of the stupendous quantities of latent heat released by condensation.

In a temperate cyclone, though condensation and latent heat also play some part, the main store of potential energy comes from the air-mass distribution in the convergence region — deep masses of cold, dense, heavy air being partly higher than warmer, rarer, lighter air masses near-by, and gravitationally urged to underrun this warmer air in order that equilibrium may be restored. While a cyclone is still a young and fairly wide warm-sector wave, its kinetic energy (wind velocities) are increasing. But after it becomes long and narrow, or 'occludes,' its kinetic energy begins to decrease.

It is a curious and interesting observed fact that cyclones along an extensive front between polar air and tropical air tend to run in families. That is to say, one cyclone forms, develops, occludes, and dies; then (or even while the first is still active) another forms along the general front to westward; then another, perhaps four or five in all. All these cyclones, of course, move generally eastward, and all may pass over or near one place, giving that place a long series of stormy-to-fair spells having an average period of five or six days. Thus the weather often seems to repeat itself at five- or six-day intervals.

WARM AND COLD FRONTS

Having surveyed the general shape of a typical temperate cyclone, we can now examine in more detail its three-dimensional structure — and that with particular reference to its surfaces of discontinuity between unlike air sectors — its 'fronts.'

As diagrammed on the 'plan view of a typical cyclone', the *warm front* (actually a wedge of cold air overlaid by warmer air) extends east and southeast from the low centre, with the warm sector advancing behind it from the southwest and the colder air retreating slowly ahead of it towards the north. As the warm air is advancing much faster northeastward than the cold air is retreating, particularly aloft, it must continually overrun the cold wedge, and steadily climb the inclined plane of the cold air's upper surface — a slope usually several hundred miles long, and several miles high at its outer part.

The *cold front*, extending southward and southwestward from the low centre, is also a wedge of cold air underlying warm air -but an active, undercutting wedge, a steeper wedge than the warm front, a wedge that is steadily advancing eastward and southeastward in such a way as to crowd out the warm air more or less violently and to thrust it aloft.

Between the warm and cold fronts in the southern part of the cyclone, and being steadily displaced and pinched out by the cold front, is the *warm sector*.

It is evident on the plan diagram of the cyclone -which might be a picture of a well-developed low as drawn on the weather map (a similar low does occur, in fact, on 'Portion of the North American Weather Map, Second Morning') — it is here evident that a large, wide area of clouds and rain extends northward and eastward from the warm front, and that a narrower area of clouds and rain occurs along and behind the cold front. Why? The answer is to be found in a vertical section through the cyclone.

The wedge of cold air is retreating slowly; the warm-moist air is overrunning this wedge rapidly by the wind arrows. As the warm air climbs the cold slope, it cools adiabatically, soon reaching its dewpoint. In the area *Nbst,* where there is still plenty of moisture in the air, it forms heavy nimbo-stratus clouds from which rain or snow steadily falls. (If the warm air is very moist and unstable, it may produce some cumulo-nimbus, heavy showers, or even thundershowers in this region.)

As the overrunning warm air ascends higher on the slope in the second *Nbst* region, it is still saturated and precipitating ice particles and droplets of water; hence it forms a wide sheet of nimbo-stratus cloud from which sprinkles light but continuous rain or snow. As the overrunning air ascends still higher, into the *Ast* region, it is still saturated, but has so little moisture left that it forms only a thin layer of alto-stratus cloud, with little rain or snow, which may evaporate before reaching the ground.

In the *Cist* region the overrunning air is now very high, quite cold, and has almost no moisture left, but it is still saturated and forms a thin layer of diffuse cirro-stratus, composed of minute ice crystals. Still higher and farther eastward, even this tenuous cloud thins out into mere wisps of cirrus perhaps seven miles above sea level.

To an observer on the ground, the approach of a typical warm front is signalized first by 'mare's-tail' cirrus clouds, 'locks of the approaching storm,' rising usually out of the southwest or west. Cirrus alone may occur as fair-weather clouds near the centre of a high. But if the cirrus thickens to cirro-stratus, which in turn thickens and lowers to alto-stratus, if 'last night the moon had a wedding ring; tonight no moon we see,' it is fairly clear indication that a warm front and its cyclone are really on the way.

Meanwhile the surface winds are mostly from the south, southeast, or east, and the pressure is gradually falling as warmer, lighter air is replacing the colder, denser air overhead. The under alto-stratus surface may lower more rapidly as myriads of ice crystals — nuclei for raindrops — begin to fall through it. As the alto-stratus lowers and thickens into nimbo-stratus, a light continuous rain (or snow) at the ground begins very gradually, and gradually becomes heavier under lowering clouds as the warm front approaches. In the region of heavy rainfall, there may be 'scud' clouds below the main cloud deck, caused by the excessively moist and perhaps turbulent atmospheric conditions. After the passage of the warm front, unless there is a cold front mixed up with it or immediately behind it, the sky usually clears partly and the rain stops, the surface air is warmer, and the pressure remains low. But over a cold water or land surface, the warm sector may be blanketed by low clouds or fog.

The wedge of cold air is advancing rapidly, underrunning the warm air. In the region *Cunb* the warm-moist air is being lifted most violently and cooled most quickly. As a result, there are heavy cumulo-nimbus clouds here, with heavy showery rain or snow, and perhaps thundershowers. (Frontal thunderstorms can occur in winter.) Behind this *Cunb* zone there is perhaps a narrow zone of rainy, chaotic nimbo-stratus, marked *Nbst* in the diagram. Farther back in the advancing cold wedge, around the region *Cu*, the cold air mass itself may be fairly moist and unstable. As a result, cumulus clouds and scattered showers may continue for some distance behind the cold front, particularly around mountains where rising currents start the unstable air upward.

As the cold-air wedge rushes southeastward, friction at the ground retards its progress in the lower levels, particularly if the ground is hilly. Hence the slope of the cold front tends to steepen. In extreme cases some of the cold air aloft may bulge out somewhat (ten or twenty miles) ahead of the cold front at the ground. In this region the temperature gradient may be very steep and the air distribution may be very unstable, resulting in violent overturning with heavy up-and-down drafts.

Where a thick, steep wedge of cold air is pushing rapidly into considerable depth of warm, moist, unstable air, the front may be marked by a continuous line of thunderstorms, and is called a 'line squall.' (Incidentally, any large thunderstorm may exhibit in miniature some of the characteristics of a cold front.) These line squalls — enormous and awesome manifestations of atmospheric violence .

To an observer on the ground, the approach of a cold front is usually signalized by a line of heavy cumulus-type clouds (probably cumulo-nimbus tops on a solid base of nimbo-stratus) approaching from the west or northwest, which may be preceded and partly screened by other clouds such as false citrus and altocumulus. Parallel banks of alto-cumulus, particularly, are the characteristic forerunners of many cold fronts. (From an airplane at high altitude the main line of cumulus tops can usually be seen better, and sometimes extends visibly for hundreds of miles.)

As the front approaches, the pressure tends to fall gradually, becoming unsteady. The heavy clouds, often dark and threatening in appearance, continue to rise out of the west, and below them may be a gray wall of rain. The pressure begins to rise sharply as colder air arrives overhead, and the south or southwest breeze is interrupted by heavy puffs out of the west or northwest, which build up into a gusty wind that may be a gale. Meanwhile the rain or snow slants downward in showers that are heavy, but usually brief.

Finally, perhaps an hour or two after frontal passage, the showers have become lighter and less frequent, the pressure is still rising, and the clouds begin to break away. But if the cold air mass is very moist and unstable, showers and cloudiness intermittently renewed may delay the final clearing for several hours, or even for a day or two, particularly around and on the windward slopes of mountains. This usually happens after cold-moist (Polar Pacific) outbreaks on our mountainous Pacific coast; and also, to some extent, on the windward (northwest) slopes of the Alleghenies.

The winds in the warm sector, ahead of the cold front, are mostly from the southwest. Usually these winds increase with the cold front's approach (and consequent crowding of warm air), and in a violent line squall, the high and gusty surface wind may shift very suddenly from southwesterly to northwesterly. But fairly often there is a relatively calm zone along the front itself, so that its passage may be marked by the gradual dying away of the warm sector southwesterly or southerly winds, a stagnant milling-about of clouds and showers in almost calm air, and then a wind, increasing gradually, out of the northwest or north.

Violent wintertime cold fronts across the northern United States usually take the form of a blizzard — a howling, cold, piercing gale out of the north or northwest, attended by blinding snow falling from the skies and blowing up from the ground. By the time the front reaches the Southern States, however, it is usually much less violent and less steep in slope — it may, in fact, be stalling and becoming a warm front. Hence its bad weather — perhaps cold rain falling through the thin wedge of below-freezing air — may take the form of an ice storm.

Clouds and showers in a summer line squall are heavy, but usually rather scattered, and limited to a narrow zone along the front itself. But in winter — at least in the central and eastern United States — the cold-front cloud-and-precipitation zone is wider, the cloud ceilings are lower, and visibility is much worse than in summer. The temperature may not drop any faster in winter with a cold-front passage, but it usually keeps on going down, and the cold is made more uncomfortable and hazardous by the howling north winds of the invading cold air mass, which may roar for a day or so after frontal passage. In the northern plains states, a violent winter cold front can drive the mercury almost visibly downward in the tube; one instance is recorded where the temperature dropped more than sixty degrees in two hours — from about fifty above at 6 A.M. to about ten below at 8! If the cold air mass is moist and unstable, snow squalls may continue intermittently for hours after the main front has gone by. And if the cold mass sweeps out over relatively warm ocean, it may raise a doubly dangerous 'steam' fog that rides masthead-high with the gale.

One of the most violent cold fronts in recent United States weather history occurred on Armistice Day, November 11, 1940.

Many duck hunters had taken advantage of the holiday to hike and row to far woodland-water places in pursuit of the fastflying birds, which were migrating south in great numbers. Suddenly, at each successive place, the cold-front line squall and blizzard — the most severe and destructive experienced this century — struck out of the northwest.

Before frontal passage there was little or no wind. The front arrived with showers of hail and a sixty- to eighty-mile-an-hour squall wind; and this bitter, blinding blow continued at gale force, with steadily falling temperature, for twenty hours. Many of the far-scattered and isolated huntes and campers died of exposure. On the Great Lakes, many steel steamships were sending out S.O.S. calls, and at least one large freighter went down in the terrible short, high lake seas with all hands to a watery grave. Damage from wind in the city of Chicago alone ran over a million dollars.

UPPER FRONTS AND OCCLUSIONS

So much for the structure of simple warm and cold fronts. What happens as the cyclone of continues its development? The cold front is moving faster than the warm front, and overtakes it, gradually pinching out the warm sector. This process, called 'occlusion,'. The dotted line here represents the occlusion — that is, the combination of warm and cold fronts where the latter has overtaken the former. The occlusion itself usually extends gradually southward as more and more of the warm sector is forced above the surface by the closing wedges of colder air.

The northern end or 'tail' of the occlusion may finally swing southward and eastward, forming a 'secondary' cold front attended by considerable bad weather. This secondary cold front or 'bent-back occlusion' of course usually begins as an upper front (explained below), but its bad-weather effects are often projected down to the surface, perhaps to such an extent that a new surface front is formed. (Secondary cold fronts can also, more rarely, form at the surface, behind the main cold front, when wind-and-temperature discontinuities exist in the cold air mass.)

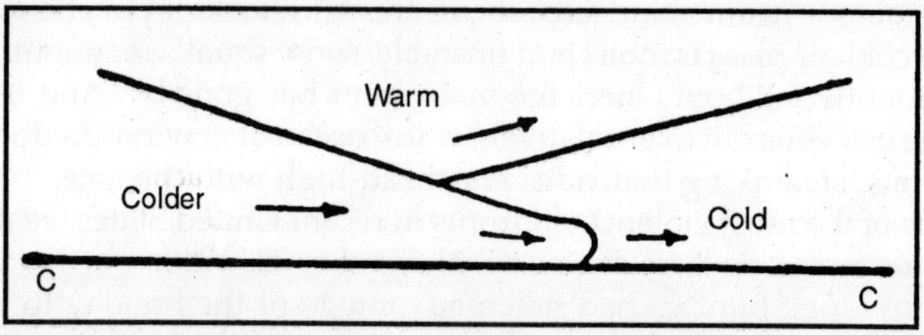

Fig. 52. Simplified Section View of a Cold Occlusion

Considered in vertical cross-section, an occluded front may take either of two general forms, depending on which of the two cold air masses (east and west of the warm sector) is the colder.

Usually the cold air behind the warm sector is colder than the cold air ahead of it. After the cold front has overtaken the warm front, obliterating the warm sector at the ground, it continues to plow along the earth's surface, lifting up in turn the less cold air beyond the warm front.

The section CC across such a 'cold-front type' occlusion, which we may call simply a *cold occlusion*. The warm sector, pinched out at the surface, exists for a time aloft. Ahead of the upper occluded warm front, on the retreating wedge of less cold air, are clouds and rain of the warm-front type; along the occlusion (surface cold front) and behind it are clouds and showers of the cold-front type. Ahead of the whole cold occlusion the surface pressure is falling; behind it, the pressure is rising.

The less frequent 'warm-front type' occlusion, or more simply named, *warm occlusion*. Here the air behind the warm sector is less cold than the air ahead of it. After pinching out the warm sector, this overtaking, less-cold air cannot displace the wedge of colder air ahead of the warm front, but rides up on it behind the warm mass. Here again the warm sector exists for a time aloft after it is obliterated at the surface, and the cloud forms of both warm and cold fronts are combined along the warm occlusion.

If the overtaking, lesscold air mass is fairly warm and very moist, and if the original warm front is moving slowly (or perhaps stalled against a mountain range), the warm-front rain may continue, or even increase, until a new cold front arrives from the west, deep enough and cold enough to plow under the preceding fronts and sweep them eastward as an occlusion of the cold type. In this (warm) type of occlusion the wedge of less-cold air, climbing the inclined plane of the more-cold air wedge ahead, is in effect a cold front at upper levels — an *upper front* that may not give much surface indication of its presence and progress. Nevertheless, it exists very definitely aloft as a trough of warmer air, and may cause considerable bad weather.

These upper fronts arising out of warm occlusions are most frequent in wintertime, when a warm front from the Pacific Ocean stalls against the high west coast mountains. The incoming cool, unstable Polar Pacific (cold-moist) air rides up over the stalled warm front and over the coastal mountains; continues on eastward over the Rockies; and perhaps crosses the Midwest, above a lower shallow layer of polar Canadian (cold-dry) air, as an upper cold front marked by an extensive high deck of alto-cumulus, alto-stratus, or strato-cumulus clouds. From this high cloud deck light rain or snow may fall.

The wind-shift accompanying this front is, of course, at high levels (around one or two miles), so that surface indications of the frontal passage are scanty. About the only surface sign, in fact — and that not too reliable — is a slowly falling barometer before frontal passage and a distinctly rising glass after frontal passage. In other words, a shallow pressure trough may mark the upper frontal position on the surface weather map.

In summer these upper fronts may not show any pronounced clouds, and their surface barometer indications may be feeble indeed. Yet if air conditions are favorable — if an extensive upper layer of warm-moist air is convectively unstable — the advancing cool wedge may trigger off sudden and seemingly inexplicable thunderstorms which appear to build rapidly downward from above.

It cannot be too strongly emphasized at this point that the cyclones, fronts, and occlusions are typical models, much simplified. Rarely do the actual fronts on the weather map fall into such simple and typical forms; if they did, physical weather analysis would be far easier than it is. In the photograph of 'Portion of the North American Weather Map, Second Morning,' for example, observe that the cold front to the southwest of the low is not a pure cold front, but an occlusion of the cold type.

It occluded another warm front out over the Pacific a couple of days before, and is still carrying some of the débris in the form of preceding stratus-type clouds. On the Pacific coast, near Oregon, another warm front is stalled against the coastal mountains while a cold front rushes in to occlude it from the west. The resulting occlusion will probably sweep inland, piling up in clouds and rain against the higher mountain ranges and doing nothing much in the relatively flat country between them.

In the East, warm fronts often bend around the Appalachians, assuming anything but typical shapes. Fronts of any type, anywhere, are often so mild they can scarcely be recognized, and even if characterized by plenty of wind and weather, they may occasionally be diffused over a zone a hundred miles or more wide. A front may appear to fade for a time, then regenerate into an active disturbance. One front may tread on the heels of another so that the cloud forms are badly confused.

Frontal clouds may occur, not in the solid masses diagrammed above, but in separate layers with clear layers between — it all depends on the temperature and moisture stratification of the air masses involved. The daily effect of solar heating and radiational cooling must also be considered: fog and lower clouds tend to 'burn off' during the morning; cumulus clouds tend to build up in the afternoon; most clouds tend to lower and fog tends to form during the night. All these vagaries combine to make the conscientious weather analyst's task anything but easy, despite all the aids of air-mass-and-front analysis.

DUST AND DELUGE AND GLAZE

Dust Storms

We have seen that individual temperate cyclones, and individual warm fronts, cold fronts, and occlusions, can all vary widely in their characteristics and their accompanying weather. Among the most spectacular of all these cyclonic and frontal vagaries are *dust storms, floods,* and *ice storms*.

Dust storms, or 'black blizzards,' have often tormented our American desert and plains regions. They were a living nightmare to farmers of the Midwestern 'Dust Bowl' from 1932 to 1939 — so much so that many abandoned their lands altogether and moved to less harassed sections of the country.

During the month of April, 1934, by way of instance, Iowa had no less than twenty-two dust-storm days. In Nebraska and Minnesota, also, fields were stripped of topsoil by the scouring dry winds. Day resembled night, and night resembled the stuffy blackness of the devil's own pit. In some districts people had to wear cloth over their faces whenever they went outdoors. Dust seeped into houses through closed windows, and over all there was a constant, suffocating heat and dryness. Staggering farmincome losses were entailed when growing grain was blown clear out of the ground.

Dust raised in the Midwest has several times traveled on high westerly upper winds a thousand miles or more to the eastern seaboard. One great dust cloud, originating near the Mississippi River on May 10, 1939, reached central Pennsylvania before dawn on the eleventh, and arrived at Boston six or seven hours later.

Most large dust storms occur in the late winter and early spring, when vegetation is generally scanty and winds are generally high. The conditions that conspire to raise any one storm are:

- A vigorous cyclone with high winds,
- Relatively dry air masses (usually P_c, T_s, or N_{PPC}) about the cyclone.

The dust storm starts most often in the boisterous winds behind a dry cold front — say in N_{PPC} air, originally from the northern Pacific and quite moist but considerably dried out by its passage over the mountains. Another likely place for dark blizzards is the warm sector of a violent cyclone, when this warm sector consists of absolutely unstable, warm-dry (tropical continental or tropical subsidence) air in which there is plenty of turbulence and gustiness.

Floods

Disastrous floods usually result from a combination of :

- A stalled cyclone with an excessively moist and unstable warm sector and excessive warm-front rain,
- Mountainand-valley country relatively bare of water-holding forest or brush.

The most terrible flood in all United States history occurred when, on a late spring afternoon in 1889, a dam across the upper part of a populated Pennsylvania valley burst during a torrential rainpour, and the raging waters roared down on the city of Johnstown. The storm rainfall over the Conemaugh Valley averaged more than six inches in a few hours — but this alone would not have spelled death for the people of Johnstown, had not the earthen dam with its ninety-foot water-head (which some rich Pittsburghers had built above Johnstown) burst at the height of the storm. No less than nine thousand people were drowned in that terrible hour.

A thorough study of eastern United States floods and their causes was made in 1936 by Professor Charles F. Brooks of Harvard University and Major Alfred H. Thiessen of the United States Army Aerological Service. Their general findings were as follows:

Great floods may occur in the eastern United States at any time of the year, for there is always an extensive warm-water surface near-by, from which great volumes of vapour may be transported; while, not far distant, throughout the year there are cold surfaces to furnish moving wedges of cold air to elevate the tropical air. The polar air masses engage the tropical masses normally every few days, and rains fall.

The great floods occur, however, only when a persistent high over the western Atlantic sends the tropical air inland for days in succession, and then only when the polar masses meet and elevate the tropical air over the same area or region for one to several days in succession or on repeated occasions. Topography plays an important rôle in determining where slowly moving fronts will stall: sometimes even minor divides are effective. Snow cover seldom contributes much to great floods, though by helping to keep cold air masses cold it may contribute to stalling and to the maintenance of the temperature contrast between the cold and the warm air masses.

Brooks and Thiessen showed that tropical Gulf air, originally having a temperature above 75° F. (24° C.), and a dewpoint above 68° F. (20° C.), contains something like *forty thousand tons of potentially liquid water per cubic mile*. If all this moisture, or even half of it, is condensed into rain above a given area (as it may be by strong frontal action), the inevitable result is a torrential downpour.

Most of the worst American floods occur in winter or early spring. From December to March is the season of heavy continuous rain, abetted by melting snows. Summer rains may be torrential, but they are more scattered and showery. Also, the ground is in a condition to absorb and hold water. As Brooks and Thiessen put it, 'To produce important floods in summer, monthly rains well in excess of 10 inches are required, whereas half as much will suffice in winter.'

A notable American flood was the overflowing of the Mississippi River during January-February, 1937. The first cause of this flood, like others in the Eastern United States, was a westward displacement of that 'centre of action' known as the Bermuda-Azores high from its usual mid-Atlantic position. The first result, starting in November, was an unusually mild and rainy winter for the whole area east of the Mississippi; and particularly, for the watersheds of the eastern tributaries of the Father of Waters. The end result was a maelstrom of rising waters, bursting levees, floating houses, drowning animals and men that moved slowly down the great river with the crest of the flood.

Any local or regional flood is of course intensified by mountains and valleys, particularly if these valleys are narrow canyons, and the towering slopes above are relatively bare of brush or forest (and most particularly, after a forest or brush fire). Every desert-wise camper in our great arid Southwest knows better than to camp, or even to park an automobile, along one of those innocent-looking dry washes that fan out from a high mountain canyon — at least in thunderstorm weather when these 'flash floods' are common. Compared with the Mississippi floods, these canyon affairs are of course small matters, but they can easily be drowning matters for any tenderfoot that they happen to catch unaware.

Somewhat similar in type was the local but violent Los Angeles flood on January 1, 1934. The scene was really laid in the fall, several weeks before the flood itself, by a large fire that burned off the normal scrubby brush cover of the mile-high mountains up back of Pasadena, Glendale, and Burbank.

It rained more or less steadily along the Southern California coast all Christmas week as a succession of occluding cyclones came in from the Pacific and Tp (warm-moist) air was continuously forced aloft by frontal action. On New Year's morning, it really began to rain; and by noon the streets of Pasadena were running inches deep in water. Later in the afternoon a flash flood from the high mountain canyons above struck the residential sections north of Glendale. Thirty people were drowned, and about five hundred houses swept away.

Ice Storms

When temperatures in the cold wedge under a warm front are below freezing, while above-freezing temperatures prevail in the wedge-climbing tropical maritime (warmmoist) air above, the falling raindrops are frozen into little round ice pellets, called *sleet*. But if the cold-air wedge is quite shallow (say less than half a mile thick), the raindrops do not have time to freeze before they hit the frozen ground.

On this cold, chilled-below-freezing ground — on houses, trees, wires, and the like — the splashing, sub-cooled drops instantly form clear ice or *glaze*. If the rain is very heavy and the ground objects are very cold, this glare ice may build out to a thickness of several inches on exposed branches and wires. This icy surface visitation may happen intensively over a wide area, in which case it is called an *ice storm* — a rapid regional formation of ice glaze in which trees and telephone poles alike come crashing down across roads and fields, completely paralyzing transportation and communication.

Ice storms can of course happen at suitable seasons all over the temperate zone. Like all other sorts of cyclonic storms, they occur frequently in the northeastern states. In midwinter, though, they mainly haunt the Southern States, because cold (polar continental) fronts which have advanced so far south usually become stalled or receding thin wedges which favour glaze formation, and the tropical air above is usually warm and moist enough to release copious rain. A recent southern ice storm was that of November 24, 1940. It resulted from a large outbreak of polar continental (cold-dry) air from Canada (which incidentally also caused a severe snowstorm in New England). The ice storm settled on Texas, where icing on wires grew to three inches, or even five inches, in diameter before the poles crashed. The paralyzing ice cap spread all over the Panhandle.

Falling ice chunks injured pedestrians. Babies were born, and operations performed, in heatless hospitals lighted only by hand flashlights. Conditions were so bad that no newspapers — not even a weather report or forecast -could be issued. South of the icing zone, near the Gulf, the overrunning southerly current of tropical maritime (warm-moist) air deluged the countryside with sixteen inches of rain.

FIVE TRILLION TONS OF MOVING AIR

Few people realise that this air ocean we breathe and move about in really has mass and weight. The actual mass or weight of air in lower and middle levels, over the area covered by a large temperate cyclone perhaps a thousand miles wide, is unbelievably enormous — something like five million million tons.

Perhaps one trillion tons of this air are hurrying northeastward in the warm sector; perhaps one or two trillion tons are idling northward or northwestward beneath the warm-front slope and cast and north of the cyclone centre; more than two trillion additional tons of cold air are rushing southeastward behind the cold front. Once you realise the essential fact of air's inescapable mass (and the ponderous weight and momentum of any large moving body of it), it is easy to understand the terrible force of gale winds and the widespread havoc that attends them.

But some temperate cyclones have winds that approach hurricane force, and the temperate cyclones are usually five or ten times as large, in total extent, as the more violent tropical whirls.

One of the greatest and fiercest of temperate-zone windstorms ithin historic times was that which struck England in November, 1703. The fullest account of it comes from Daniel Defoe. For generations after it was spoken of as 'the Great Storm,' and no other ordinary storm, however fierce, was allowed to usurp its title. The cyclone centre crossed eastward over the very waist of Britain.

From the relatively small area of gale winds, one might suspect that it was really a tropical hurricane gone astray, like the freak New England hurricanes of 1815 and 1938 — but the evidence, at this late date, is not complete enough to be positive. Anyhow, the damage and carnage in England, and particularly over its surrounding narrow seas, was enormous. Thousands of houses were laid flat on land, and their hapless tenants killed or maimed. Many of those who ventured outside were felled by flying chimney pots or roof fragments or tree trunks. There is no space here to recount all the tragic wrecks that befell at sea, where some 8000 lives were lost.

The earth-regions where temperate-cyclone gales roar wildest are, in general, those middle latitudes marking the average position of the main polar air-tropical air front. And these gales are always worst along the eastern shores of a large ocean, where the full, frictionless sweep of the prevailing westerlies finds boisterous play.

In the southern hemisphere, around lat. 40° S., there is little land to bar the round-the-world sweep of these planetary winds. Hence the 'roaring forties' are notorious among seafaring men everywhere. Nineteenth-century square-riggers bound around the Cape for Australia could really 'make knots' — fifteen or more each hour, faster than many present-day freight steamers. But ships and crews took stern punishment from seas that reared mountain-high astern. Fortunately a good ship's stern would always lift, at the last moment, to clear these onrushing rollers. But the mere sight of them was so terrifying to a novice that ship's boys, on their first voyage out, were forbidden ever to look astern.

Other famous places for howling gales, rolling seas, and skyshooting breakers are Land's End, England, and the various headlands and small islands along the west coast of Ireland. Here gales sometimes blow — at least in passing gusts and squalls — at shrieking speeds well over a hundred miles per hour; and the breakers along those rocky coasts are something to turn a seaman's hair white.

Winds and seas of the same terrifying sort are not uncommon along our own American North Pacific coast. In February, 1939, the American freighter *Timber Rush*, clearing the notorious Columbia River Bar outside Portland, Oregon, plowed through breakers fifty feet high — great breakers which swept another able steel steamship onto the treacherous sands that same day. On January 29, 1921, the Pacific weather station at North Head, Washington, recorded more than 96 m.p.h. of wind for five minutes, and a sustained gust of 114 m.p.h. for one minute.

Far down to southward, near the relatively calm horse latitudes, I have often watched thirty-foot breakers and hundred-foot spray-geysers bursting over the rocky headlands around Cape Conception (usually misnamed Point Conception), where Northern California changes abruptly to Southern California.

SUNSHINE AFTER STORM

One of the pleasantest things about the great cyclones of the temperate zone, and particularly our North American cyclones, is the clearing after storm, the sunshine after shadow, the breaking away of the rain clouds into blue sky (perhaps with a resplendent rainbow or two) that often attends the passage of a marked front, warm or cold.

The passing of a warm front may bring a sudden (but probably temporary) clearing, together with unseasonable yet welcome balminess in the fresh tropical air of the warm sector.

But a cold-front passage is usually more sudden, more violent, and more final as a forerunner of fine weather to come. Below the latitudes or without the seasons of blinding blizzard and perishing cold, there is something keenly stimulating and joyous about the clean, keen northerly or northwesterly gale on which a new polar air mass drives over land or sea.

Chapter 8

Weather Instruments

Through earlier and less well-organized ages, when community life was more precarious than now, anxious people used to ask, 'Watchman, what of the night?' A modern query, repeated more quietly but on a far grander scale, is, 'Watchman, what of the weather?' For the weather watchmen of today are spread over all inhabited continents and most navigable oceans.

Like the community watchmen of old, these weather men scan the sky for clouds, storms, and other portents. But they also read and interpret various instruments that tell far more about some phases of the weather than human senses alone can divine. What are these weather instruments? How do they work? What do their indications and records really mean?

HEAT AND TEMPERATURE MEAN

If we could override natural laws and look into any solid substance, such as a piece of iron, with a microscope magnifying a few million times, we could perhaps see that the iron is not really solid at all. Actually it is composed of myriads of molecules and atoms, the latter being made up of protons and electrons revolving in complicated orbits.

Even the molecules, themselves minute multiple 'solar systems' revolving with incredible rapidity, are never at rest; they are continually shooting about like erratic bullets, bumping into each other, rebounding, and shooting away in new directions.

Various animals and insects respond to temperature by greater activity or less; even a cricket is a not-too-reliable thermoscope, indicating warmth by chirping more frequently. A swarm of bees might respond to heat by flying faster and buzzing louder; if it grew colder, the bees might buzz more slowly, less loudly, until, if it grew cold enough, they would stop buzzing altogether.

The swarm of molecules in a piece of iron responds to temperature in much the same way. As the temperature falls, the iron molecules vibrate more slowly, until at *absolute zero* (the coldest possible temperature — minus 460 degrees Fahrenheit or minus 273 degrees Centigrade) they cease vibrating entirely. If the temperature rises, they vibrate faster and faster, bumping each other more and more violently. As a result of this more violent bumping, the apparently solid piece of iron grows larger, expands.

If the temperature goes high enough, the molecules pound each other so hard that they break out of their fixed orbits, begin slipping easily past one another, and the iron melts into liquid. If the temperature goes still higher, the more rapidly vibrating molecules move farther apart, and the liquid iron vaporizes into gas. Many substances respond to temperature by changing state in the same way. Water, for example, is solid at cold winter temperatures below the 'freezing point' (+32° F. or 0° C. at sea level), liquid at ordinary temperatures, and gaseous at temperatures above the 'boiling point' (+212° F. or +100° C.).

The faster the molecules move, the more energy they have, and this energy is called heat. Many people and many writings overlook the important difference between heat and temperature. Heat is the total energy of all the vibrating molecules. Temperature is the measure of their average speed of vibration.

THERMOMETERS AND AIR TEMPERATURE

Liquid Thermometers

Of all weather instruments, the ordinary thermometer is probably the simplest and most efficient.It consists essentially of a small glass bulb filled with some suitable liquid and connected to a glass tube of minute internal diameter, into which the liquid extends to a variable distance, depending on the temperature of the bulb.

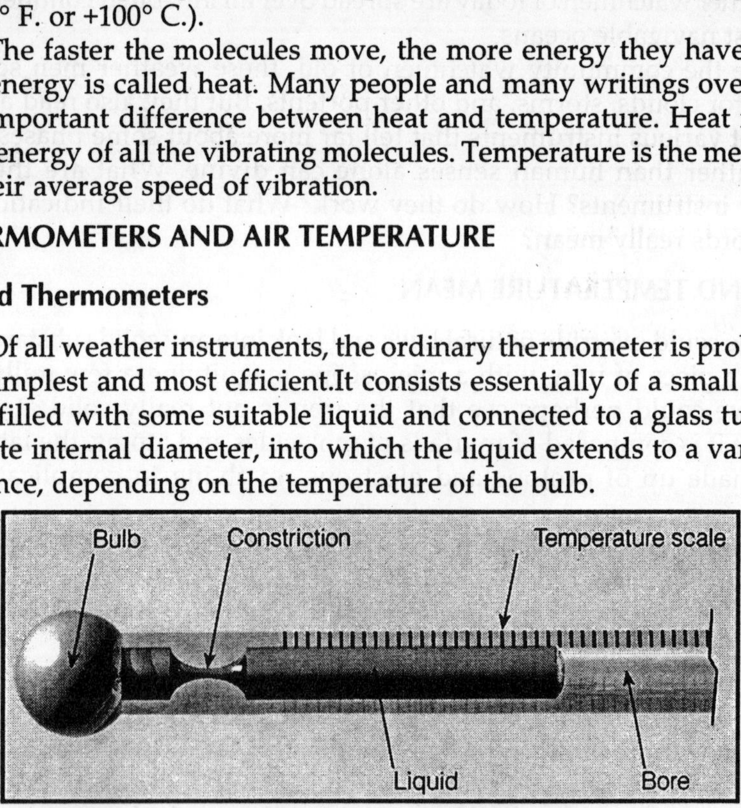

Fig. Liquid Thermometers

The liquid most commonly used is mercury, which does well enough for any temperature between its freezing-point (-38° F. or -39°

C.) and its boiling-point (+680° F. or +360° C.). For very low atmospheric temperatures alcohol (which freezes at -202° F. or -130° C.) is used, and alcohol colored red or black is coming into increasing use in ordinary thermometers because it is cheaper than mercury and easier to see in the capillary tube. But the mercury thermometer, which responds more quickly to changes in temperature, is used for most serious aerological work.

Thermometer Scales

The two reference temperatures in common use are the freezing-point of water and the boiling-point of water at sea level. Based on these reference temperatures, the Centigrade scale is the simplest and most logical; its zero is the freezing-point and its 100 is the boiling-point. This Centigrade scale is commonly used in most European countries, and scientifically used the world over.

The Absolute temperature scale is logically developed from the Centigrade scale; 0° C. is +273° A., and 0° A. is -273° C. To convert any Centigrade temperature to Absolute, therefore, simply add 773. Zero on the Absolute scale is, as mentioned above, the coldest possible temperature, at which molecular motion ceases; liquid helium evaporated into a vacuum gives a temperature which approaches absolute zero by a degree or two.

Rather illogical, but commonly used in all English-speaking countries, is the Fahrenheit scale. Perhaps Fahrenheit thought that a mixture of ice and ammonium chloride was the coldest thing possible; perhaps he thought the same of a cold wave that caught him in Iceland; at any rate, he chose one or the other as his zero. The range between freezing and boiling he divided into 180 degrees, possibly because his dividing machine happened to be built that way; as a result, the freezing-point came out at +32° and the boiling-point, +212°.

The Fahrenheit scale does have one very important advantage, though — ordinary air temperatures in ordinary places are all positive, and within the range 0° to 100°; and for ordinary atmospheric temperature measurements, the Fahrenheit degree (only five-ninths of the Centigrade degree in size) is more convenient.

In general, the Centigrade scale is better for scientific use, while the Fahrenheit scale is better for lay use. Still another scale, fortunately in very limited usage, is the Riaumur. But three scales, surely, are enough. In this book we shall meet only Fahrenheit, Centigrade, and Absolute.

Measuring Air Temperature

Reading a thermometer means, to many people, simply looking

at the scale. (On an accurate instrument the scale is engraved directly on the glass stem to avoid slippage, and it should of course be viewed perpendicularly to avoid parallax errors.) Such a reading, while giving the exact temperature of the thermometer bulb, may tell very little about the temperature of the surrounding air. This free-air temperature is what every weather man is really interested in, and it is not so easy to measure accurately as might appear at first glance. Most air thermometers in common use are often in error by five or ten degrees, if not more.

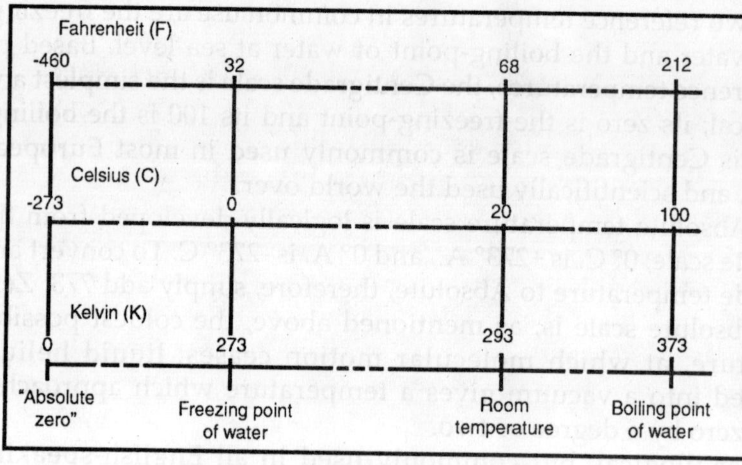

Fig. Fahrenheit Thermometer Scales

One of the cardinal rules, frequently violated by laymen, is that a stationary thermometer should never be exposed to direct sunlight. But it should, by all means, be exposed to some circulation of the free air. Also, the observer should not stay near the thermometer any longer than necessary, lest it be warmed by his body-radiation. And if a thermometer is moved to a new location, it should be allowed a few minutes to respond to the new temperature before being read.

Three general schemes of thermometer exposure are used by aerologists in their efforts to determine true air temperature. The first is hanging the thermometer from a chain or swiveled handle, and whirling it rapidly about by hand for several minutes before reading it, so that conduction of heat between the thermometer bulb and the air has a chance to nullify any errors introduced by radiation.

It should always whirl to windward of the observer's body and other near-by objects; and preferably, though not necessarily, in his or some other shadow. With a good instrument, air temperature can be measured in this way to within one half degree Fahrenheit, even though the thermometer is swung in bright sunlight.

The second scheme, a refinement of the same general principle, is the mechanically ventilated thermometer. The thermometer itself (particularly the bulb) is enclosed in one or two protective shells of non-radiating bright metal.

Air is drawn through a tube past the thermometer bulb, either by a clockwork-driven fan or by some form of rubber-bulb aspirator. A well-made, ventilated thermometer of the Assmann, clockwork-fan type is very accurate — properly operated, it will read air temperature correctly to within one tenth degree Fahrenheit.

Finally, a thermometer can be exposed to the free air by placing it in an 'instrument shelter,' which may also house various other weather instruments. The shelter is usually a cubical box between one foot and three feet on a side, having a double, ventilated top and latticed sides. The whole shelter is painted white or aluminum to minimize radiation; it is mounted on stilts at least several feet above the ground or the top of a building, and oriented so that its latticed door opens to the north.

The thermometer is hung within the shelter near the centre and well clear of the sides. Or better, it can be mounted on a revolving spindle so that it can be whirled inside the shelter by means of an outside crank. Or a small crank-operated fan can be arranged to blow on the thermometer bulb.

Maximum and Minimum Thermometers

The maximum thermometer is nothing more than a standard mercury instrument in which the capillary tube is constricted near the bulb to such an extent that, though the expanding mercury is forced out into the tube, it cannot easily retreat back into the bulb. The common clinical thermometer is simply a small maximum thermometer having an enlarged scale in the body-temperature range. In weather work, the instrument is used to record the highest temperature of the day. The minimum thermometer, used for indicating the lowest air temperature that has occurred during the night, employs clear or light-colored alcohol as its expanding liquid. When the alcohol column contracts, a small, spool-shaped indicator is pulled towards the bulb by surface tension, remaining always in the alcohol; when the alcohol column expands, the indicator sticks in place, indicating the lowest temperature reached.

There is also a combination maximum-and-minimum thermometer (Six's thermometer), containing both mercury and alcohol in a U-shaped tube together with small steel indicators that are reset by the pull of a magnet. While not accurate enough for serious aerological measurements, a Six thermometer will indicate both maximum and minimum temperature since last setting within a degree or two, which suffices for many purposes.

Dial Thermometers and Thermographs

Some types of thermometers indicate temperature by the motion of a revolving hand around a clock-like dial. The actuating element of these instruments is, in some cases, a bimetallic strip consisting of two thin, curved pieces of dissimilar metal, soldered or riveted together and clamped at one end.

As the temperature changes, the two metals expand unequally and the curvature of the strip changes; its free end is connected to the temperature-indicating needle by means of magnifying levers. In other cases, the actuating element is a flat, curved metal tube containing alcohol or some other liquid, which also changes its curvature in respo nse to temperature changes.

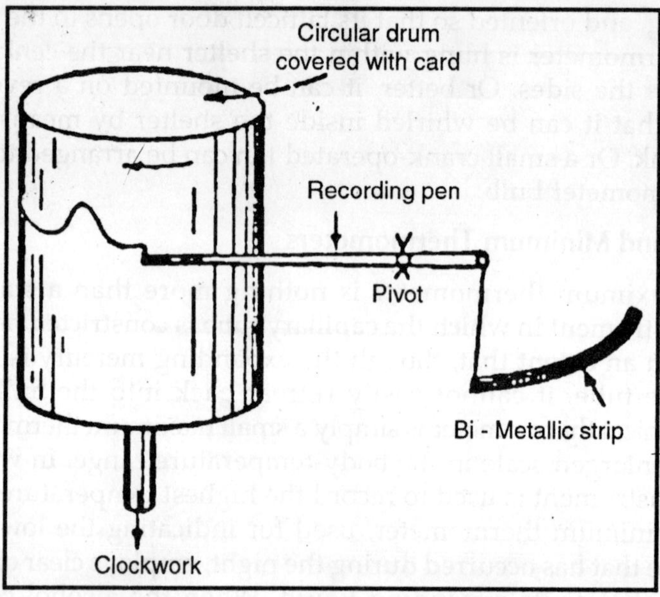

Fig. Elements of Bimetallic Thermograph

Either a bimetallic strip or a liquid-filled tube can be used as the actuating element of a thermograph, or recording thermometer. Instead of an indicating needle above a circular scale, the actuating unit and its magnifying levers serve to move a long arm having at its tip a small inked pen that presses on a drum revolved by clockwork.

As the drum revolves, the pen traces a record of the temperature and its changes on a paper sheet, fixed to the drum, which has been printed with horizontal lines corresponding to degrees of temperature and vertical lines corresponding to time intervals of an hour or two. The drum is usually adjusted to make one complete revolution in a week. Other forms of thermographs have variously shaped and scaled record sheets moved by clockwork in various ways, but their general principle is the same. As thermograph elements and their magnifying levers are not inherently very accurate, thermographs used in serious aerological work are checked against mercury thermometers every day or so.

Remote-Indicating Thermometers

Two types of thermometers serve to indicate outside temperature to an observer within a building. In the capillary type of instrument, the temperature bulb is connected by an insulated capillary tube, usually not more than twenty feet or so long, to the indicating dial. These thermometers are subject to creep and some other difficulties, but do indicate a remote temperature continuously and directly. In the electric-resistance type of instrument, the sensitive temperature element is connected to the indicating device by three wires, which may be several hundred feet long.

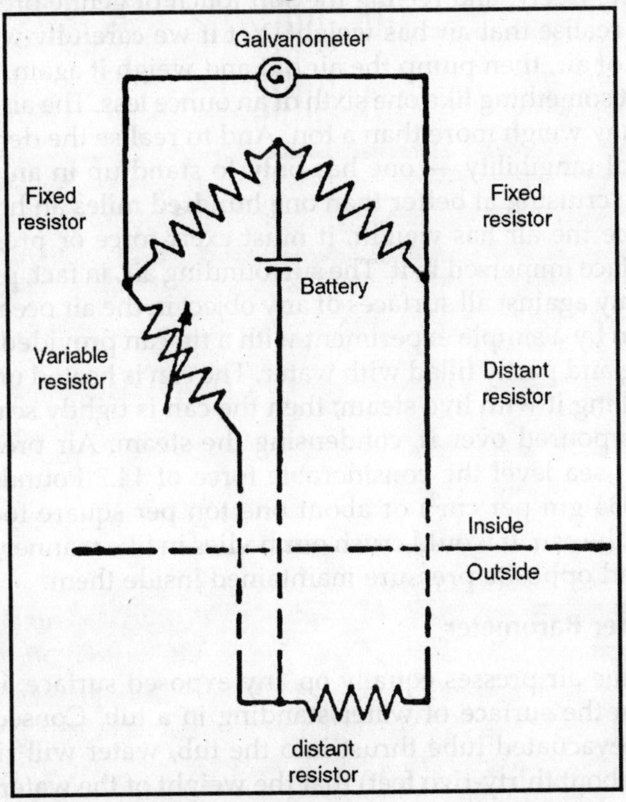

Fig. Essentials of an Electric Resistance Thermometer

The underlying principle here is the Wheatstone bridge, by means of which currents in electric circuits are balanced in order to find the value of an unknown resistance in terms of a known one. The unknown in this case is a coil of fine wire in the remote sensitive element, whose re-sistance varies with the remote temperature. The dial of the adjustable balancing resistance, which the observer turns while watching a galvanometer and bringing its reading to zero, is calibrated to read directly in degrees of temperature. This electrical instrument can also be used, with more complicated apparatus, for recording distant temperatures on a chart. Another type of electric-resistance thermometer employs a capillary tube filled with electrolyte as the temperature-sensitive resistance.

BAROMETERS AND ATMOSPHERIC PRESSURE

Atmospheric Pressure

When we are moving slowly about on the bottom of the air ocean, breathing freely, and feeling the soft touch of gentle breezes, it is not easy to realise that air has weight. Yet if we carefully weigh a gallon jug full of air, then pump the air out and weigh it again, it will tip the scales at something like one sixth of an ounce less. The air in a spacious room may weigh more than a ton. And to realise the density of air — its actual tangibility — one has only to stand up in an open-cockpit airplane cruising at better than one hundred miles an hour.

Since the air has weight, it must exert force or pressure against any surface immersed in it. The surrounding air, in fact, presses almost uniformly against all surfaces of any object in the air ocean. This effect is shown by a simple experiment with a tin can provided with a small opening and partly filled with water. The can is heated until the water boils, filling it with live steam; then the can is tightly scaled and cold water is poured over it, condensing the steam. Air pressure, which exerts at sea level the considerable force of 14.7 Pounds per square inch (1034 gm per cm^2) or about one ton per square foot, promptly crushes the can. It would crush our bodies in like manner, were not an equal and opposite pressure maintained inside them.

The Water Barometer

As the air presses equally on any exposed surface, it must press down on the surface of water standing in a tub. Consequently, in a vertical evacuated tube thrust into the tub, water will rise to such a height (about thirty-two feet) that the weight of the water column just balances the air pressure on an equivalent area of the water surface in the tub. In other words, the weight of the water column balances the weight of an air column of like width, or cross-sectional area, extending to the top of the atmosphere. The water barometer is unwieldy, and evaporation and other factors make it inconvenient to use. It has, nevertheless, the advantage of being very sensitive to pressure changes, for even a minor atmospheric disturbance causes the water column to sink or rise an inch or two.

Mercury Barometers

The mercury barometer, of more convenient dimensions, is exactly similar to the water barometer in principle. The mercury type is used today for all accurate pressure measurements, and appears in several different forms. Simplest and most versatile is the ordinary manometer, a U-shaped or J-shaped glass tube partly filled with mercury. This device is used for a great variety of pressure measurements in physics and engineering.

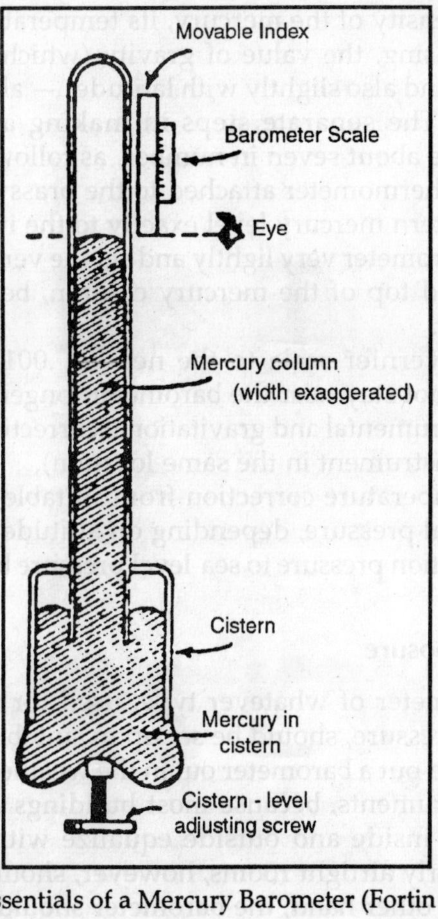

Fig. Essentials of a Mercury Barometer (Fortin Type)

The standard weather barometer, of somewhat different design, consists of one small, straight glass tube about thirty-five inches long, dipping into a cistern of mercury about the size of an inkwell, the whole being enclosed in a suitable casing of heavy brass and hung vertically in a protective box.

A movable index near the top of the mercury tube, carefully adjusted until its lower end appears just to touch the rounded top of the mercury column, measures on a vernier scale the exact height of the mercury column above the mercury level in the cistern.

Measuring Pressure

A carefully made pressure measurement with a good Fortin mercurial barometer should be accurate to one thousandth of an inch of mercury (one thirtieth of a millibar). The process involves several different steps, no one of which is suited to careless minds or bungling fingers. The density of the mercury, its temperature, the temperature of the brass casing, the value of gravity (which varies considerably with altitude, and also slightly with latitude) — all these must be taken into account. The separate steps in making a complete pressure observation are about seven in number, as follows:

Read the thermometer attached to the brass casing.

Set the cistern mercury level exactly to the ivory point.

Tap the barometer very lightly and set the vernier indicator exactly on the rounded top of the mercury column, being careful to avoid parallax errors.

Read the vernier scale to the nearest .001 inch (or tenth of a millibar). (Do not stay near the barometer longer than necessary.)

Apply instrumental and gravitational corrections (always the same for the same instrument in the same location).

Apply temperature correction from suitable table to find station pressure (actual pressure, depending on altitude of station).

Reduce station pressure to sea-level pressure by consulting suitable table.

Barometer Exposure

Any barometer of whatever type, used for the measurement of atmospheric pressure, should be set up in suitable surroundings. It is not necessary to put a barometer out in the weather with thermometers and other instruments, because most buildings are leaky enough so that pressures inside and outside equalize within a few moments. Airtight or nearly airtight rooms, however, should be avoided.

But on the other hand, the barometer should not be located near open windows, particularly if it is windy outside, on account of dynamic pressure effects. Finally, the barometer should not be placed where it will be subjected to much vibration. If vibration of its surroundings is unavoidable, it should be floated on sponge rubber, rubber shock cords, or flexible springs — a principle that applies to the mounting of any delicate instrument.

Pressure Units

Like temperature scales, pressure units are still in the dark ages of unnecessary duplication. Until a few years ago most European countries expressed atmospheric pressures in millimeters of mercury, while the British Empire and North America expressed them in inches of mercury. Now, how- ever, most countries have gone to millibars, a true pressure unit that is the international standard for serious aerological work. In the United States, millibars are used by the Weather Bureau and most professional aerologists, while most amateurs and laymen still use inches. Most American barometers, of course, still read in inches of mercury.

Some day, perhaps, all atmospheric pressures will be measured in millibars. The precise relation between the three units is such that 1013 millibars equals 760 millimeters of mercury equals 29.91 inches of mercury equals 14.7 pounds per square inch. Average sea-level pressure, in other words, is very close to 30.00 inches and quite close to 1000 millibars. The millibar is a true unit of pressure which fits smoothly into modern physics, for one millibar denotes a pressure or force of one thousand dynes per square centimeter.

ANEROIDS AND PRESSURE CHANGES

Aneroid Barometers

These widely used instruments, named from the Greek words for 'not wet,' appeared during the middle of the nineteenth century. A small vacuum chamber of thin, corrugated nickel alloy, perhaps about the size and shape of a small watch, is mounted in the barometer case, and its round faces are held out against atmospheric pressure by an external steel spring.

If the air pressure decreases, the spring pulls the faces of the vacuum chamber out slightly; if the pressure increases, it pushes the faces in slightly against the force of the spring. One face of the vacuum chamber is fixed rigidly to the barometer case; the other is connected by means of multiplying levers to a pressure-indicating needle that moves around a cir- cular scale. The best aneroids are 'compensated' for temperature changes by admitting a little air into the vacuum chamber; others are intended for use at ordinary room temperatures only. A good aneroid barometer should read pressure accurately to about one hundredth of an inch (one third of a millibar) if properly adjusted and used. But the aneroid should, if possible, be frequently compared with the more accurate mercurial barometer or some other standard reference, as it is subject to various errors. The circular scale is usually marked off in twentieths or fiftieths of an inch, which enables one to note the position of the needle to the nearest hundredth. (Millibar scales are usually marked off in unit millibars, and can perhaps be read to the nearest fifth of a millibar, though the nearest millibar should do for most purposes.)

Reading an Aneroid

This operation is much easier than reading a mercurial barometer. Simply tap the instrument *very* lightly and read the pressure by the indicating needle to the nearest hundredth of an inch (or millibar). An index needle that can be set by hand is provided on many weather aneroids, but it is not ordinarily needed by weather observers, who record times and observed pressures in writing. Similarly the misleading words 'Rain,' 'Change,' and 'Fair,' which adorn most lay aneroids, are paid scant attention by trained aerologists and should be left off the scale entirely.

Pressure vs. Altitude

Greater changes in pressure than those occasioned by the weather can of course be observed at any time by the simple expedient of changing one's elevation on or above the earth's surface. Specialized types of aneroid barometers, such as the surveying aneroid and the airplane altimeter, are hence used to indicate height above sea level.

These instruments can never be absolutely accurate, because their scales are based on a 'standard atmosphere,' assuming an average distribution of temperature and an absence of moisture which, in nature, seldom or never exist. Also, pressure altimeters take no account of atmospheric pressure changes that may have occurred since they were last set. But actual conditions are usually such that pressure altimeters give a close approximation of the true elevation, which does well enough for many purposes.

Barographs

Atmospheric pressure at any fixed elevation is always changing, in temperate latitudes, under a variety of causative factors. Most large weather-observing stations keep an automatic record of these hourto-hour and day-to-day pressure changes by means of a barograph. This instrument includes either a single vacuum chamber similar to that used in an aneroid or, more likely, a multiple arrangement of vacuum chambers somewhat resembling an accordion. Instead of being connected to an indicating needle, the vacuum element actuates through suitable levers a recording pen, which traces the pressure record on a revolving drum or moving card similar to that used in a thermograph.

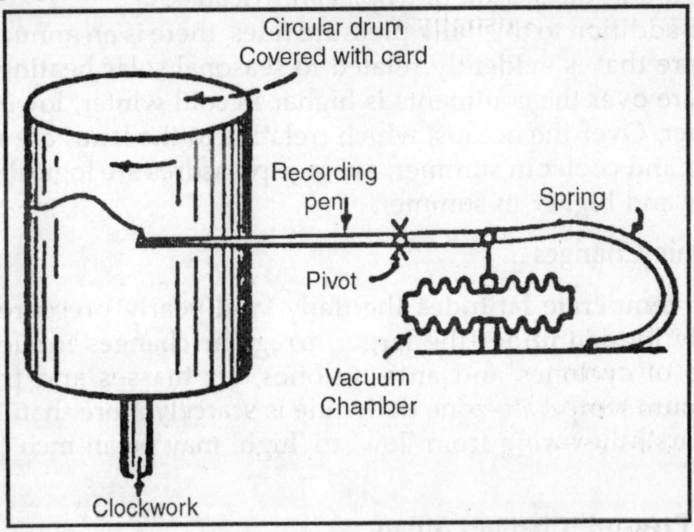

Fig. Essentials of an Aneroid Barograph

Pressure Tides

Barometric observations show that, in addition to the larger irregular sweep of cyclonic 'lows' and anticyclonic 'highs,' there are small daily tides in atmospheric pressure, strongest near the equator (where their total rise and fall may reach .01 inch or 3 millibars), but felt to some extent over most of the earth. The high-pressure tide occurs simultaneously along a meridian at about 10 A.M. and 10 P.M., local time; low pressure comes at about 4 P.M. and 4 A.M.

As these pressure tides are repeated at the same solar times day after day, as they are most marked near the equator, and more marked in summer than in winter, it is beyond doubt that they are caused in some way by the sun. Certainly the low-pressure tide at 4 P.M., the warmest hour, results from air aloft being heated and rarefied. The reasons for the remaining tides are not well understood, but may depend on such things as delayed heating, cooling, and ionization at upper levels, together with symmetrical pressure waves on opposite sides of the earth.

It is known, for example, that increasing ionization by solar ultraviolet rays at altitudes above sixty miles results in increasing pressure at the ground; this might have some connection with the 10 A.M. pressure maximum. As a practical way of eliminating the regular pressure tide from cyclonic or other weather pressure disturbances, barometers should be read at 7 A.M., 1 P.M., 7 P.M., and 1 A.M., local time. This precaution is important within the tropics, and particularly important in the season of tropical hurricanes.

In addition to the daily pressure tides, there is an annual march of pressure that is evidently related to seasonal solar heating. Average pressure over the continents is higher in cold winter, lower in warm summer. Over the oceans, which (relative to the land) are warmer in winter and cooler in summer, average pressures are logically lower in winter and higher in summer.

Cyclonic Changes

In temperate latitudes the daily 'and yearly pressure tides are usually hidden under the larger, irregular changes induced by the sweep of cyclones and anti-cyclones, air masses and fronts. The maximum temperate-zone daily tide is scarcely more than .06 inch (2 millibars); the swing from 'low' to 'high' may be an inch (35 mb) or more.[2]

What Pressure Changes Mean

In essence, what does a 'low' barometer reading or a 'high' barometer reading really mean? In a general way the 'low' reading means that the air overhead is relatively rare and light, probably because it is relatively warm; or that there is a dynamic spreading-out of air aloft. The 'high' reading means that the air overhead is relatively dense and heavy, probably because it is relatively cold; or that there is a dynamic piling-up of air overhead.

It must be remembered, also, that conditions *through the whole height of the atmosphere* are being indicated by the barometer, so that the relative 'coldness' associated with high pressure, say, may perhaps be occurring in the stratosphere rather than in lower levels. In a very general way, then, a rising barometer means that denser (probably colder) air is piling up to greater depths overhead, displacing rarer air. A falling barometer means that denser air overhead is being replaced by rarer (probably warmer) air, which is often moister air accompanied by clouds and rain.

A region where denser air is being replaced by rarer air aloft is, therefore, a region of low pressure on the surface. If isobars, or lines of equal surface pressure, are drawn on a map of this region, they will indicate an area of low pressure, or 'low,' in the same way that contour lines on a topographic map indicate a valley.

On the other hand, a region where rarer air is being replaced by denser air aloft is a region of high pressure on the surface. Isobars drawn on a map of this region will therefore indicate a high-pressure area, or 'high,' in the same way that contour lines on a topographic map indicate a hill.

In temperate latitudes these lows and highs drift slowly across the weather map, usually (but not always) in a general west-cast direction. Hence at any fixed station on the ground, a barometer needle (or barograph pen) begins to fall if a low approaches it from the west. As the low centre passes over the station, the barometer needle reaches its lowest reading, and thereafter rises as the low recedes eastward.

If a high follows on the heels of the low, the barometer needle continues to rise until the high centre has passed, after which it falls again, perhaps with the approach of another low. Highs and lows which pass near the observing station, though not centrally over it, cause similar (though less marked) fluctuations of the barometer.

MOISTURE AND ITS MEASUREMENT

Water Vapour in the Air

Some water is always present in the lower levels of the air ocean, even over the Sahara Desert. For the most part, it takes the invisible form of water vapour which is thoroughly mixed with the air. Not until the amount of water vapour becomes large enough, or the temperature falls low enough, does this invisible vapour condense into the visible form of dew, fog, or clouds. The amount of water vapour in the outside air varies from a mere trace up to about 4 per cent by volume or 2½ per cent by weight.

Specific Humidity

The most useful of moisture measures is the 'specific humidity' — that is, the actual mass of water vapour present in a given mass of air. Specific humidity is usually expressed in grams of water vapour per kilogram of air, and its value varies, under actual diverse weather conditions, from less than 1 g/kg in dry, cold air up to more than 20 g/kg in warm, moist air.

Specific humidity is the best all-around measure of atmospheric moisture because it is 'conservative,' tending to re- main constant in spite of certain variable conditions. Other measures of humidity change greatly with changes of pressure and temperature, but provided no water is evaporated into vapour and no water vapour condenses into liquid form, specific humidity always remains the same.

Saturation Specific Humidity

For every temperature, there is a 'saturation specific humidity' — that is, a maximum amount of water vapour that the air or space we are concerned with (at that temperature) will hold in the invisible form of vapour. At 32° F. or 0° C., for example, the saturation specific humidity is 3.5 g/kg; at 68° F. or 20° C. it is 14.6 g/kg. When moisture increases (or temperature decreases) until the saturation specific humidity for the existing temperature is reached, condensation begins (if certain contributory factors are favorable), and the invisible moisture comes out in the form of visible dew, fog, or clouds.

Relative Humidity

The most widely used measure of moisture in the air is 'relative humidity.' The relative humidity is simply a ratio expressed in per cent — the ratio of the actual, existing specific humidity to the possible, saturation specific humidity at the existing temperature. In the outside air, relative humidity values range from perhaps twenty per cent or so over deserts to eighty or ninety per cent over or near warm oceans. But inside a house in winter, when we heat dry, cold air up to livable temperatures, the relative humidity may fall well below twenty per cent, and the air we breathe may be drier than the harmattan blast of the Sahara.

Dewpoint

Another common measure of moisture in the air, and a very useful one, is the 'dewpoint.' The dewpoint is the temperature at which the presently existing air will, if it cools sufficiently, become saturated with moisture. A more correct though somewhat more involved way of defining the dewpoint is to say that it is the temperature having a saturation specific humidity equal to the existing specific humidity.

For in any ordinarily moist air the existing specific humidity (which remains constant until condensation starts) must be the saturation specific humidity for some lower temperature; when this lower temperature is reached, dew begins to appear on exposed surfaces and water vapour begins in other ways to condense into liquid form. As an ordinary measure of outside-air moisture, dewpoint is better than relative humidity because it is less affected by temperature and other changes.

From our definitions of these common measures of moisture and the relations between them, it is now clear that any one of three conditions indicates that condensation of water vapour into visible form is imminent —

- The nearness of the specific humidity to the saturation specific humidity,
- The nearness of the dewpoint to the temperature,
- The nearness of the relative humidity to one hundred per cent. These relations will be worth remembering when we come to the formation of fogs and clouds.

Hair Hygrometers

The simplest apparatus for indicating moisture in the air, and in various forms the most widely used, is human hair. A suitably prepared, single hair about a yard long, hanging in still air and stretched taut by a small weight, should indicate relative humidity as accurately as the most ornamental and costly hair hygrometers in use today. A commercial hygrometer usually includes an actuating element consisting of several hairs in parallel, often arranged somewhat like the strings of a banjo. The lengthening and shortening of the hairs is geared by means of magnifying pulleys and levers to a needle that travels over a circular scale. The scale of a good instrument is usually marked in one per cent divisions but, unfortunately, this does not mean that the instrument is accurate.

Used under favorable conditions and frequently compared and corrected with a psychrometer, a good hair hygrometer may be trusted to within five or ten per cent. But in the absence of such precautions it may read almost anything. At humidities below twenty per cent, the hairs are likely to dry out and may give very faulty indications. And subjected to moderate cold — say below +15° F. or - 10° C. — the instrument is practically useless.

Electric Hygrometer

An entirely different principle is used in the electric hygrometer. Two coils of wire, insulated from each other but closely parallel-wound on a glass tube, are covered with some moisture-sensitive material such as lithium chloride. As relative humidity varies, so varies the electrical resistance between the two coils. By means of suitable balancing circuits the electric hygrometer can be designed to indicate humidity at a considerable distance from the sensitive element.

Hygrographs

A hygrograph outwardly resembles both the thermograph and the barograph. Its construction is also exactly similar, except for the actuating element of hairs used either singly or in multiple. A good hygrograph is usually built more carefully than a hygrometer, and is hence likely to be more accurate under favorable conditions, though still leaving a good deal to be desired. But the hair hygrometric principle does indicate the relative humidity continuously and directly, without further ado. Most other hygrometric instruments are more complicated.

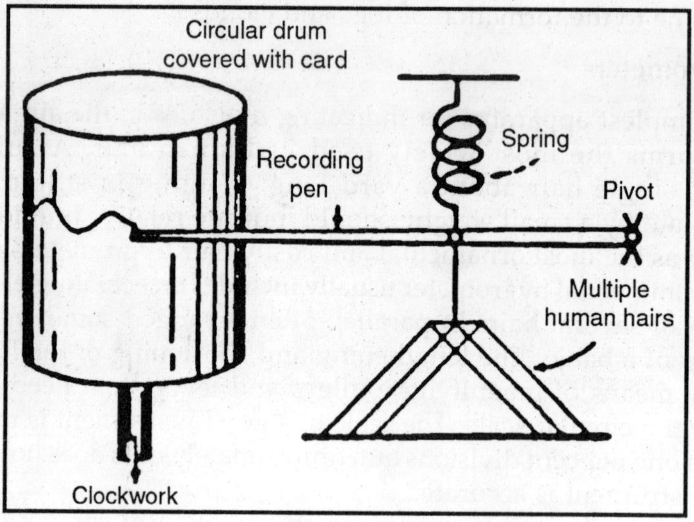

Fig. Essentials of a Hair Hygrograph

Dewpoint Hygrometers

If both the temperature and the relative humidity are known, the dewpoint can be easily computed or found from tables. But it is also possible to find the dewpoint directly by means of suitable instruments. The simplest of these is nothing more than a familiar ice-water pitcher, preferably made of bright metal. When it is brought out on the lawn on a hot summer day, it does not remain bright long. Very soon a film of dew forms on the bright surface, and if we could read the temperature of the pitcher at the moment when the dew forms, we should have the exact dewpoint.

More convenient and businesslike than the ice-water pitcher is a small, bright-metal tube filled with ether, into which dips a thermometer — the standard dewpoint hygrometer. When air is bubbled through the ether, it cools by rapid evaporation, and the sinking mercury in the thermometer tube is watched closely until, at the moment when dew forms on the tube, it indicates the dewpoint. To make an accurate measurement, one has to run the temperature of the ether back and forth through the dewpoint several times.

The Psychrometer

The most accurate of all practical humidity- measuring instruments, and the most generally satisfactory, is the psychrometer. This consists of two identical mercury thermometers, mounted parallel to each other and close together. The bulb of one thermometer is enclosed in a close-fitting sleeve or wick of cloth, wet with water, evaporation from which causes the 'wet bulb' to indicate a somewhat lower temperature than the 'dry bulb.'

The higher the relative humidity and the closer the dewpoint to the temperature, the closer together do the wet and dry thermometers read. By noting their respective readings and consulting a table, both the relative humidity and the dewpoint can be found in a moment. At moderate temperatures and fairly high humidities, the dewpoint is usually something like twice as far below the dry bulb as the wet bulb.

Given dry-bulb temperature 55° F. and wet-bulb temperature 50° F. at sea level, for example, the dewpoint is 45° F. (and the relative humidity 70 per cent). But at lower temperatures the factor is greater than 2, at higher temperatures perhaps 1½ or less; and tables or special slide rules should always be used for accurate determinations.

In a widely used form of pyschrometer, both thermometers are mounted vertically on a stationary board, with the wick that encloses the wet bulb dipping into a small well of water. But unless such a device is exposed to freely circulating air it cannot be at all accurate — for in stagnant air evaporation from the wick is blanketed by a cushion of moist air around it. Also, the wick may get clogged with dirt or precipitates in the course of a few days, failing to evaporate the water properly anyway.

As used for regular weather observations, the psychrometer usually takes the form of two accurate mercury thermometers mounted on a narrow strip of metal; and is often attached to a sling, so that it can be easily swung around in a circle like the whirling thermometer. The dry-bulb reading is, in fact, a whirled-thermometer reading, and gives quite accurately the temperature of the free air. The observer should whirl the instrument in the shade if possible, and also to windward of him, so that his body warmth and moisture may not affect it. The wet bulb should be covered with loosely woven, washed muslin, frequently renewed. Just before the observation this bulb is dipped in clean water having about the temperature of the outside air. Then the psychrometer is whirled for a moment, the dry and wet temperatures are read, the instrument is whirled again, read again, and so on until the correct dry-bulb reading and the lowest wet-bulb reading are obtained. If the temperature is below freezing, the wet bulb must be allowed to freeze completely before starting the observation.

This same type of psychrometer — two good scale-on-stem thermometers fixed to a narrow metal frame — is often mounted in a latticed instrument shelter where it can be whirled by an outside crank, or where air can be blown past it by a fan. There is also the ventilated psychrometer (similar to the ventilated thermometer), enclosed in one or more bright-metal cases and including some sort of blowing device. These instruments are rather complex and expensive in their refined form, but probably give the most accurate humidity determinations of all.

RAIN AND SNOW AND OTHER MATTERS

Rain Gauges

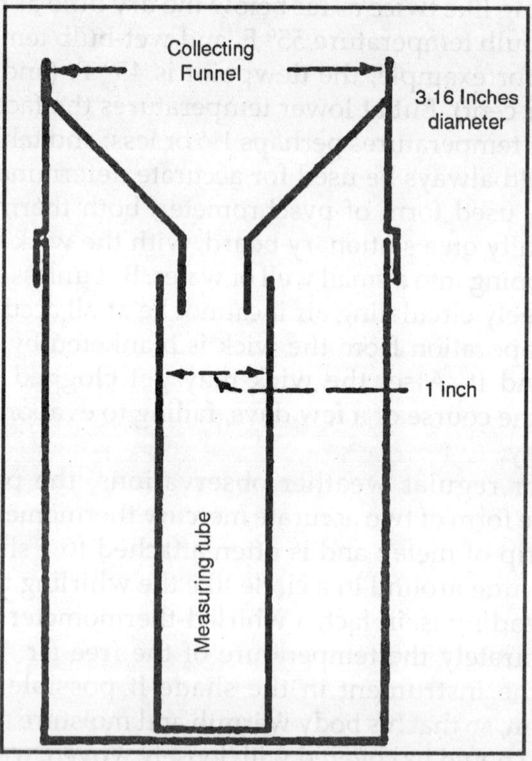

Fig. Cross-Section of a Simple Rain Gauge

Rainfall was the first of the weather elements to be accurately measured, and the ordinary rain gauge is today one of the simplest of weather instruments. In regions of heavy rainfall, a fair gauge might be made from nothing more than a round can of uniform width open at the top to receive the rain. Then, after the rain stopped, one could push a ruler into the collected water and note the height to which the ruler was wetted; this height would indicate the amount of rainfall directly in inches and fractions. To correspond with regular rainfall observations, however, the water height should be measured in inches and hundredths (or millimeters, in the metric system).

In order to register conveniently in such small units the actual rainfall must be 'magnified'; the rain is collected by some sort of funnel, whence it runs into a smaller tube for measurement.

If the rainfall is to be magnified exactly ten times, the round collecting funnel must be 3.16 times as wide as the round measuring tube — that is, the cross-sectional area of the funnel top must be ten times the area of the tube. The complete gauge includes the upper funnel-shaped basin, a large cylinder over which it fits snugly, the inner measuring tube, and a special ruler graduated in 'inches' each ten inches long and divided into hundredths.

Another type of rain gauge, used at larger observing stations, records electrically each hundredth of an inch of rain as it falls. The water collected by a large upper funnel pours into one side of a small divided bucket of special shape, pivoted at its centre. When this side of the bucket is filled, it tips downward, closing the electric circuit, emptying itself, and throwing the other side of the bucket under the funnel.

Snowfall

The depth of snowfall is usually measured by pushing a measuring-stick down through the snow at several places and taking the average of the measured depths. A location free from drifting must be chosen, of course, if the measurements are to mean anything. The water equivalent of a fall of snow can be found by pushing a cylindrical tube down into the snow so as to get a sample, which is then melted in the tube. As snow varies greatly in texture, its water equivalent also varies. But as a broad average, ten inches of snow may be considered somewhere nearly equal to an inch of water.

Cloud Observations

Various cloud observations, made in the course of many years, have told much about the motions of the upper air. First of all, an observer notes the amount of cloudiness present in the sky. He estimates this cloudiness amount in tenths of the whole sky covered — scattered clouds may cover two or three tenths of it, broken clouds seven or eight tenths. He also notes the kind of clouds present in the sky. Finally, the observer notes the direction from which the clouds are moving and forms some idea of their speed. Ordinarily this is done by simply watching the clouds drift, preferably from a viewpoint directly beneath the four compass arms of a wind vane or the cross-arms of a convenient electric-line pole. Or more elaborately, the cloud drift (and hence the upper winds) can be more or less accurately measured by means of nephoscopes, vertical telescopes, or camera obscuras.

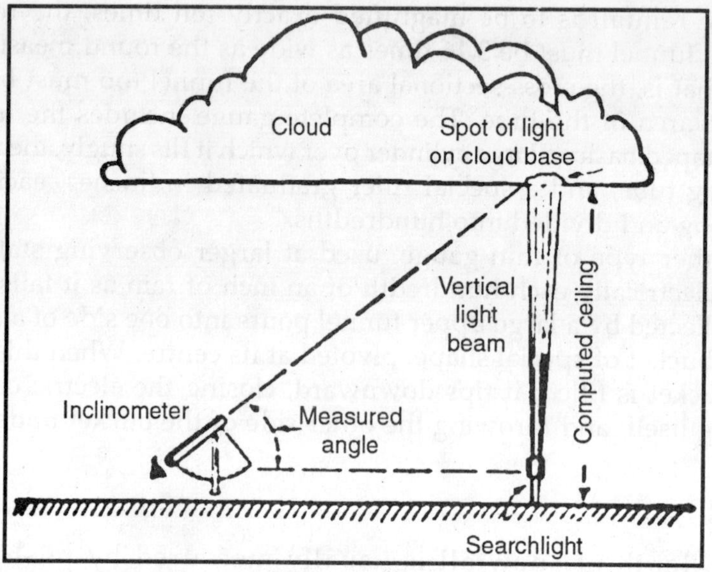

Fig. Principle of the Ceiling Light

The height of cloud bases, or 'ceiling,' is a very important measurement in aviation aerology. Usually it is merely guessed at, with very approximate results unless several reference marks such as high towers and mountain-peaks of known height are conveniently scattered about. At higher levels the best guide is a knowledge of cloud types and their ordinary altitudes. For more accurate ceiling measurements in the daytime, a small rubber balloon, inflated with hydrogen to a measured 'free lift' that has been found to give it a known rise rate, is released and timed during its ascent to the cloud base.

At night a simpler and more certain method is available -the ceiling light. A powerful searchlight, usually five hundred or one thousand feet away from the observer, projects its beam vertically on the cloud base. The observer sights an inclinometer on what usually appears as a rather nebulous spot of light in the clouds, reads the angle, and finds the ceiling height in feet from a table. The ordinary short-base ceiling light is quite accurate for ceilings below three thousand feet or so, where accuracy is most needed. Much above three thousand feet, with ordinary equipment, the light spot becomes rather vague and the angle too critical. A powerful mercury-arc searchlight at the end of a thousand-yard baseline might give these higher ceilings more accurately.

Though rather complicated visibility-meters are available, practically all present measurements of visibility, also important in aviation, are made by eye. Landmarks available by day and lights available by night — buildings, trees, hillocks, and so on — at various known distances in all directions from a weather station, are listed in a table and used by the observer.

The Complete Observation

Weather watchmen scattered over the world keep close check on every important weather development in civilized regions. A complete 'surface' weather observation, as performed by any one observer at any one time, consists of the following instrumental measurements:

Amount of rain or snow since last measurement
Temperature
Dewpoint
Wind direction and velocity
Barometric pressure (corrected and reduced to sea level)
Extent and character of pressure change in past three hours

The observer also notes by eye:

Clouds — including proportion of sky covered, kinds of clouds, their direction of motion, and the heights of their bases

Horizontal visibility, and any obstacles thereto such as fog, haze, etc.

State of the weather — raining, snowing, thunderstorm, etc.

Any unusual meteorological phenomena.

Thus it is scarcely correct to say, any longer, that nobody does anything about the weather. Nobody does much about changing it, but many people are reporting it and not a few are analyzing it, with an accuracy and completeness never before attained in the world's history 1. Some idea of the average (standard atmosphere) relation between pressure and altitude in the lower levels is given by the following table:

Table. Average Variation Of Atmospheric Pressure With Height

Altitude (in feet)	Altitude (in meters)	Pressure (in inches of mercury)	Pressure (in millimeters of mercury)	Pressure (in millibars)
Sea level	Sea level	About 30.00	About 760	About 1015
2,000	610	27.80	705	945
5,000	1524	24.90	635	845
10,000	3048	20.60	525	695
15,000	4572	16.90	430	570
20,000	60.96	13.75	350	465

Table. Sea-level Pressures Likely To Be Observed In Different Weather Conditions

Type of weather	Pressure (in inches of mercury)	Pressure (in millimeters of mercury)	Pressure (in millibars)
Tropical hurricane or typhoon	About 27.00 to 28.50	About 685 to 725	About 915 to 965
Extreme temperate cyclone or 'low'	29.00	735	980
Ordinary temperate cyclone	29.50	750	1000
Average sea-level pressure	30.00	760	1015
Ordinary anti-cyclone or 'high'	30.50	775	1035
Extreme anti-cyclone	31.00	785	1050

Chapter 9

Effective Mountain Sounding

While the brilliant new Norwegian technique of physical weather analysis was sweeping American aerological circles in the nineteen-thirties, there arose a great hue and cry about regarding weather as 'three-dimensional.' Some of this uproar was justified, for many professional and amateur weather men had lapsed into the inert habit of forecasting weather from nicely rounded isobars of surface pressure as drawn on a two-dimensional map. But included in air-mass-and-front analysis publicity there was also the implication that, before the advent of the Norwegian technique, American old-timers in the Weather Bureau and elsewhere had been scarcely conscious that the air ocean possessed any such quality as height or depth.

This implication was unjustified. And coming from some over-publicized weather analyst scarcely out of college, it naturally aroused the fulminating wrath of every seasoned aerologist who had been sending up sounding kites and balloons for the past twenty or thirty years. Moreover, it aroused considerable unnecessary doubt and malice towards the whole technique of air-mass-and-front analysis itself. The oldest and simplest way of getting up into the air — and perhaps still the healthiest and pleasantest — is by climbing a mountain.

Moreover you can actually make, on a mountain, worth-while aerial soundings of a sort. Strangely enough, however, effective mountain sounding came later in history than balloon sounding. Anyone who has tried to signal by flag or flashlight to or from a mountain-peak partly or wholly hidden in clouds will know why. Mountain sounding, at least in the form of continuously operated and fully useful mountain-top weather observatories, had to wait on the invention of the electric telegraph.

MOUNTAIN SOUNDING

The first regular mountain weather observatory in the world was established in 1871 on a sub-arctic peak that pokes up six thousand-odd feet into the storm-harassed skies of northern New England —

Mount Washington. The Army Signal Service (which was also the Weather Bureau in those days) at first furnished only a telegraph line over which observations could be reported into the national wire networks. But within a year the mountain-top reports were proving so useful that the Signal Service took over the observatory and put it on a more or less permanent basis.

Two years later, encouraged by the success of Mount Washington Observatory, the Signal Service established the world's highest weather station at 14,110 feet altitude on the summit of Pike's Peak in Colorado.

Pike's Peak is a majestic mountain, and so far the only threemile summit in America that you can ascend quickly and easily by mechanical transport. Though it may be torrid in the cities below Pike's Peak, as you gain altitude on the slopes of this great mountain the air becomes gradually cooler.

At the summit there is something that closely approaches winter. It may snow on short notice, even in midsummer. At the top, also, you are likely to feel mildly dizzy, or to notice a slight headache, because the lungs are only getting half their normal amount of oxygen at each breath. Those who stay over will probably not be able to eat very much, or to sleep very well, especially the first day.

Several years ago I spent twenty-four hours on the summit of Pike's Peak, writing in my journal as follows:

Up here on the summit, this now-become-cool July afternoon, the weather is threatening. Above, and away in all directions towards the wide horizon, are towering clouds darkened in places by sheaves of falling rain that evaporates before reaching the ground. Here and there a lightning bolt spits abruptly downward from some dark cloud base level with our lofty perch.

Past midnight, and the clouds have cleared away above the mountain. But they still hang in low, black fringes around the horizon ring. The domed summit slopes away downward in all directions, like the curve of some miniature planet, some asteroid in dim space. Thirty miles to the southeast gleam the patterned street lights of Pueblo; seventy miles northward a broader lightcluster marks the position of Denver. Here on the dark summit, with nearly half the earth's atmosphere below, man is closer to the splendors of the heavens.

The stars shine out in unwonted brilliance. So many new, bright stars are sprinkled around the sky that you can scarcely recognize the familiar constellations. Every few seconds a brisk morning meteor streaks hurriedly across the zenith, coming from somewhere near the Pole Star. A busy breeze eddies around the summit, and it is cold — somewhere near the freezing point. But out on the Kansas plains, over there in the dark east, heat-weary people are sweltering on blanketless beds. Above them gleams Venus, rising two hours ahead of the sun. In the northeast a solitary thunderstorm is fitfully flashing somewhat below our level, doubtless noisy and windy enough to the beings below it, but a mere toy spitfire from here.

Faint daylight is flooding up out of the east as Old Sol prepares to scorch the prairies anew. A dozen overheated autos and twoscore super-cooled people are up for the sunrise. Small lakes in the Colorado Springs basin gleam like puddles of molten silver in the dawn. Thin clouds low in the east are ablaze. Now a line of intense light appears above the Kansas plains three hundred miles away, and the edge of a fiery disk pokes up over the rim of the world, lighting the ragged clouds with scarlet and orange and yellow. In no time the glowing ball, peering over the horizon, shoots level, yellow-red rays at the hushed summit, the old stone house, and the little crowd of early watchers huddling into the depths of their overcoats and blankets.

Mount Washington and Pike's Peak were maintained by the Government as regular weather stations, day in and day out, summer and winter, for about seventeen years. The mountainobservatory idea spread to Europe, where peak stations were established in the Alps and other mountainous districts. But by 1890, for devious bureaucratic reasons, the two American mountain-top stations had been abandoned.

In 1932 Mount Washington Observatory was re-established by Harvard University and other scientific institutions. The present station includes instruments designed to study the problems of mountain blizzards, to measure high-level hurricanes that sometimes blow at more than two hundred miles an hour, and to probe air masses and fronts by means of ultra-short-wave radio.

This renaissance of the Mount Washington station started a general reawakening of interest in mountain observatories all over the world. Because of topographic effects on wind, heating and cooling, mountain aerological measurements are not as representative of general free-air conditions as measurements taken with kites, airplanes, or balloons. But the mountain station has one decided advantage: it is always there at one fixed altitude, by darkness or daylight, blow hot or blow cold.

It would be a splendid thing if we had more first-rate mountain observatories in America. Suitable peaks, besides Mount Washington and Pike's Peak, might be: Mount Mitchell in North Carolina (6711 feet — highest of the Appalachians); Mount Whitney in California (14,502 feet — highest in the forty-eight states); and Mount Rainier in Washington State (poking up 14,408 feet into the North Pacific storm tracks). An ideal summit observatory might take the form of a metaland-concrete building topped by a dome of plate glass or quartz from which observers could watch storms and stars in all-year high-altitude comfort.

Some sort of powerful snow-removing equipment (such as a hot-water fountain) would of course be needed. For dependable and quick all-year transport, the peak might be connected to lower levels by electric trains running in a tunnel bored at a slant up through the rocky heart of the mountain.

What with the automobile roads that now wind up many mountains, a rough-and-ready sort of three-dimensional moun tain-sounding method is available to anyone (very approximate and amateurish, of course, yet sometimes worth while). All you need for a simple temperature sounding (aside from a surveying altimeter to indicate the successive 500-foot levels in the first place) is a sun-shielded thermometer plastered by two suction cups on the front of the automobile windshield. On most steep roads, of course, a 'downward' sounding can be made more quickly than the upward climb, and amounts to the same thing. Best accuracy (or least effect from slope-heating and slope-cooling) is obtained around sunset or sunrise, which are also the best times for mountain views.

Here, for example, is the record of an automobile sounding made on a fair-sized peak in California. We left the summit (4000 feet) at 7.06 A.M. (eleven minutes after sunrise), and arrived at the base (600-foot altitude) at 7.30 A.M.

Table. Mountain Auttomobile Sounding

Altitude	Temperature (F.)	Remarks
600	50°	Sky, clear; calm
1000	51°	
1500	53°	Temperature inversion
2000	50°	Temperature decrease
2500	45°	Steep temperature decrease
3000	48°	Temperature inversion
3500	42°	Steep temperature decrease
4000	44°	Wind north, 5-10 m.p.h.

In general, this sounding confirms the average physical truth that temperature decreases with increasing altitude. But there were two layers; one between 1000 and 1500 feet, the other between 2500 and 3000 feet, where on this particular morning the temperature actually increased with altitude. These 'temperature inversions' are a common phenomena of late night and early morning. Between 1500 and 2000 feet there was a temperature decrease of 3° F. in 500 feet (usually expressed as a 'lapse rate' of -6° F/1000 feet). Between 3000 and 3500 feet the lapse rate was steep indeed −12° F/1000 feet — a lapse rate only possible in still air.

These increases and decreases of temperature with altitude mean a great deal to an aerologist, for they tell him how 'stable' or 'unstable' the air layers in question are. This stability or instability, in turn, indicates a great deal about the likelihood of vertical air currents, clouds, rain, fog, and other weather phenomena.

These mountain-sounding figures must be taken with more than a grain of salt, for they are subject to various undesirable inaccuracies. (But at sunrise these inaccuracies are not very great.) To get more accurate evidence of temperature and humidity, stability and instability, in the upper air, we must turn to such devices as kites, airplanes, and balloons.

KITES AND CAPTIVE BALLOONS

By about 1900, after free-ballooning enthusiasm had begun to wane, kites provided the simplest and most reliable means of aerial sounding up to altitudes of around two miles. The kite itself had developed into an open box-like affair some seven feet on a side, capable of lifting perhaps twenty pounds. Four or five kites, or occasionally more, were connected in tandem to several miles of thin but strong steel wire which was unwound from a power-reel driven by steam or electricity.

The upper kite carried a clockwork-driven aerograph, which recorded values of temperature, pressure, humidity, and wind velocity against time on a slowly turning drum. From the recorded traces made respectively by the thermograph, hygrograph, and wind-recording pens, the variation of temperature, humidity, and wind with increasing altitude could be determined.

Kite sounding with recording instruments really originated in America at Blue Hill Observatory in Massachusetts. Later it was extended to other American stations by the Weather Bureau, and European aerologists also adopted it. Some of the kitesounding sagas around the turn of the century were almost as exciting, in their own way, as the airplane-sounding sagas that were to come a generation later.

One evening at Blue Hill a line of five kites was out on five miles of wire when a gale suddenly arose. As the steam winch started to haul in, the kite wire snapped at the drum and the whole tandem whirled away into the gusty darkness, dragging the steel wire across country. Within a few minutes it had shortcircuited and disabled a trolley line to the accompaniment of blinding flashes of electric fire. Most of the wire came to light next morning when it snarled around the wheels of a railroad train and stopped it for nearly an hour.

The Weather Bureau kite-sounding station at Broken Arrow, Oklahoma, also had its share of excitement. One day a long tandem ascended through a 'warm front' marked by northeasterly winds at lower levels and southwesterly gales aloft. When the wire broke near the reel, the lower kites pulled against the upper ones in such a way that the whole tandem held its altitude and drifted away into the next state.

Weather Bureau kite men also lost kites by 'icing' — the bane of present wintertime airplane flying. And one day a long kite tandem was caught in the violent up-and-down drafts of a marked 'cold front' or line squall, which beat them down despite the pull of the five-horsepower motor reeling in at full speed. Up to the last minute the ground crew worked frantically at the reel despite frequent flashes of lightning. Finally a large bolt struck the kite wire and instantaneously melted its lower portion into drops of hot metal. But because the reel was well grounded, the crew escaped injury.

As kites can only be used efficiently when there is a wind of more than ten miles per hour, captive-balloon sounding was developed for use in still air and very light breezes. Wire and reel were much the same as those used with kites. The round balloon itself, hydrogen-filled and perhaps ten feet in diameter, with the aerograph slung several feet below it, would lift around forty pounds of wire. In light airs a single balloon could reach heights of two or three miles, and two balloons in tandem might occasionally reach as high as four miles.

Military kite balloons, teardrop-shaped and fitted with stabilizing fins or their tails, combine some of the advantages of both kites and captive balloons. The 'observation' type is too large and expensive for ordinary aerological use, but the British 'barrage' type, about fifty feet long — or perhaps a somewhat smaller 'blimp' -might be ideal.

Kites and captive balloons have one advantage over most other methods of sounding: they give a cross-section of the air above one definite place. A sounding balloon, or even a sounding airplane, may drift a hundred miles or more in the course of its climb. But with the widespread development of American aviation during the nineteen-twenties, steel wires stretching skyward became too much of a flying hazard to be tolerated — except where, as in the British balloon barrages of the Second World War, they were deliberately erected for the express purpose of keeping low-flying enemy bombers away from vital and vulnerable areas such as London. A somewhat similar captiveballoon project is the use of bright orange, mirror-studded balloons by movie studios, to keep noisy airplanes away from their sound recordings.

There still remains, however, a reasonably safe method (and a simple and inexpensive one) of captive-balloon sounding to low altitudes in light winds. This technique, which we developed at Bolling Field in 1936, gave temperature and humidity data up to one or two thousand feet above the surface — data that was very useful in analyzing certain weather situations, such as a dense fog blanketing a busy airport and necessarily halting its urgent flying schedule.

Of course such an airport fog situation, in addition to worrying operations and weather officers over the vexed question of when it will clear, blocks off the more usual methods of upper-wind finding and aerial sounding that would in part supply the answer. Hence a single determination of temperature aloft, a rough measurement of upper gradient wind, and definitely locating the top of the fog layer may all be well worth while.

An oversized pilot balloon, inflated to a diameter of about three feet, carries the sounding instrument aloft at the end of a light but fairly strong fishing line unwound from a geared reel. The sounding instrument is preferably a miniature aerograph of the Jaumotte type, although it may be no more than a high-lag thermometer arranged to hold its reading for some time, or a black-bulb maximum thermometer that will show a high reading when it emerges in sunlight above the fog. The angle of the string gives a rough idea of wind speed and direction, allowances being made for the catenary droop, and the gradient-to-surface curvature, of the string itself.[1]

Using an ordinary inexpensive rubber balloon, this device is practicable for light winds — i.e., flat calm up to fifteen or twenty miles per hour. Balloon, sounding instrument, and string are all of such light construction that any airplane which happens to strike them would cut harmlessly through without damage to itself.

Airplane Sounding

Aviation's phenomenal growth during and after the First World War provided a great impetus to aerial sounding — in two ways. First, it created more demand for upper-air data than had ever existed before. Second, it provided the quickest and simplest instrument (and up to moderate altitudes the least expensive and most efficient one) for making the desired soundings.

Any small sport airplane out of the thousands owned or rented by American private pilots can be used to make a perfectly good temperature-aloft sounding by simply tying a suitable thermometer securely to some outboard part of the plane.

The strut thermometer itself should of course be placed where it can easily be seen and read by the pilot or observer. Though an ordinary fifty-horsepower light plane climbs at scarcely more than two hundred and fifty feet per minute (even solo, and considerably slower, of course, with two people), this slow climb rate aids accuracy by eliminating thermometer lag.

Like our amateurish mountain automobile sounding, this light-airplane sounding reveals a temperature inversion in the lower levels. These temperature inversions are a very common occurrence during early morning hours, particularly in winter and spring; and they are caused by the night-time cooling and settling of the lower-level air, which in turn is cooled mainly by the radiational cooling of the earth's surface. In hilly or mountainous country, this lower-level air cooling is abetted by the downslope drainage of cooler (and denser) air; and you often find, at night, even small hollows several degrees cooler than the hillocks near-by.

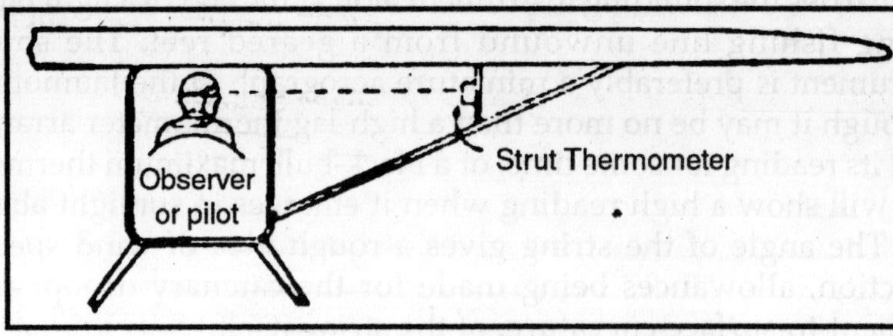

Fig. Strut Thermometer on Light Airphane

The result of all this night-time cooling, which reaches a maximum just before sunrise, is an accumulation of colder (and denser) air, liquidlike, in the lower levels, with warmer air above. The air in these temperature inversions is very 'stable' — it resists vertical currents. The inversion usually persists until it is wiped out by intense ground heating under the rising morning sun. It happened also on this particular morning that the air below fifteen hundred feet was practically all at the same temperature — constituting an 'isothermal layer.'

Above three thousand feet the temperature begins to drop off rapidly, which means, in effect, that this air layer is 'unstable' — a term to be explained at some length farther along. It is worth noting here, however, that in this unstable layer, the airplane bounced around in considerable 'bumpiness' or turbulence.

For a professional sort of airplane sounding — say one made by the United States Army in order that military weather forecasters can keep track of air masses and fronts, or a sounding to determine 'ballistic density' aloft for heavy artillery firing -we must step up both the power and the expense. Instead of fifty horsepower there are five hundred or more in the motor, and the plane itself is strong enough to withstand quick climbs, turns, acrobatics and dives. The maximum climb rate might be two thousand feet per minute or more, but in practice it is usually held down to 1000 ft/min or less because of instrument lag.

A specially designed sounding instrument is used — a streamlined aluminum case about the size of a football, firmly clamped to a steel strut that projects out one and one half feet (far enough to clear the zone of engine heat) from the fuselage just opposite the observer's cockpit. Inside the aluminum case, in position to be easily and quickly read by the observer, are an accurate centigrade thermometer and a high-grade (accurate within five to ten per cent) hair hygrometer.

In 1936 we designed and built such an instrument at Bolling Field, and used it for many quick soundings up to twelve thousand feet or so. We even did loops, spins, and vertical power dives with the sounding instrument clamped in place on the plane, in order to prove its strength and rigidity.

Table. Military Airplane Sounding

Altitude (ft.)	Pressure (mb.)	Temperature (-°C. [°F.])	Lapse Rate (°F.)	Relative Humidity (%)
20 (airport)	1015	+26° (+79°)	--	95
1,000	970	23° (74°)	-5°	88
2,000	930	34° (76°)	+2°	85
3,000	895	22° (73°)	-3°	82
4,000	870	21° (70°)	-3°	80
5,000	788	19° (67°)	-3°	78
6,000	807	18° (65°)	-2°	76
7,000	776	15° (59°)	-6°	81
8,000	745	13° (56°)	-3°	80
9,000	723	11° (52°)	-4°	83
10,000	695	10° (50°)	-2°	80
11,000	668	8° (46°)	-4°	78
12,000	640	6° (43°)	-3°	75

Instead of noting altitude directly in this sort of airplane sounding, the observer notes pressure in millibars on the dial of a barometer mounted on the instrument board in his cockpit. The pilot takes off and climbs steadily at 300 to 500 ft/min. Reaching five hundred feet above the airport, the pilot levels off into horizontal flight for fifteen seconds, during which time the observer reads his barometer. Then, as the pilot noses up into the next 500-foot climb, the observer reads his thermometer and his hygrometer.

On a pad strapped to his thigh the observer notes down pressure (mb.), temperature (°C.), and relative humidity (%) at each 500-foot level. If the ground aerologists are in a hurry for these data, the observer can radiophone them down immediately from each 500-foot level. At slow-climb rate, allowing for instrument lag, a sounding to twelve thousand feet altitude can be made in half to three quarters of an hour.

This sounding shows several interesting facts. Here again is the familiar temperature inversion in lower levels, this time between 1000 and 2000 feet, even though it is deep midsummer in the torrid August climate of Washington. Above 2000 feet the temperature decreases with altitude in the way that we should expect — at about -3° F. per 1000 feet. *This particular 'lapse rate' (-3° F/1000 ft) is the normal lapse rate for average free- air conditions all over the world.* In temperate latitudes the average summer lapse rate is somewhat steeper than this, and the winter rate is somewhat less steep because of intense surface cooling during the long winter nights.

Between 6000 and 7000 feet this sounding shows a very steep lapse rate of -6° F/1000 ft, and as a matter of observation, we noticed considerable bumpiness or turbulence in this layer. The sounding reveals also that, even on a torrid midsummer day, you can find real and refreshing coolness by climbing seven or eight thousand feet in an airplane. At the top of the sounding, the 43-degree temperature felt like the North Pole after a sweltering night in Washington. Here at 12,000 feet, too, the new-risen sun was blindingly bright, and shone full on the thermometer scale with such an intense iridescence.

Table. Winter Airplane Sounding

Altitude (ft.)	Temperature (F.)	Lapse Rate	Relative Humidity
Surface	36°	--	76 -- raining
2,000	26°	-5°	100 in clouds -- snowing
4,000	21°	-2½°	100 in clouds
6,000	18°	-1½°	100 in clouds
8,000	19°	+½°	100 in clouds
10,000	14°	-2½°	98 above clouds
12,000	6°	-4°	82.

This military-airplane sounding may be called a typical American summer sounding, penetrating through two and one half miles of mostly summer tropical maritime (warm, moist) air. Now let us examine, by way of contrast, a typical winter-storm sounding taken at Boston on one of those cloudy, rainy, snowy days that make New England weather what it is:

In this case the cyclonic storm winds, extending from the ground up to high levels with increasing force, have prevented the formation of any stable temperature inversion in lower levels by reason of their turbulence and the consequent violent atmospheric mixing at all levels. It is worth noting here, also, that the lapse rate in the clouds is around -2½° F/1000 ft -and we shall find that (for sufficient scientific reasons) such a lapse rate always exists in active clouds and storms at moderate altitudes.

The air masses revealed by this sounding are, at lower levels, transitional polar continental (cold, dry) air; and at upper levels, overrunning transitional winter tropical maritime (warm, moist) air that has been drawn by the upper cyclonic circulation all the way up from the Gulf of Mexico. One point that should be stressed about our sample soundings is that they were taken in relatively moist, warm air, and hence show relative humidities considerably higher than average. Drier air masses, at higher levels particularly, would show relative humidities of fifty per cent, thirty per cent, or less.

Temperature soundings are now often taken by outgoing air liners as they climb to cruising levels after leaving a terminal. As the air liner climbs steadily to ten thousand feet or so, the pilot notes the temperature at each thousand-foot level and radios these data back to his air-line aerologists.

There is another sort of 'airplane' sounding, of somewhat dubious scientific value, but eminently spectacular. That is parachute sounding — where a parachute jumper, falling clear at one hundred and fifty miles an hour or so on a delayed jump, or drifting downward at a mere fifteen miles an hour once his 'chute has opened, takes along a thermometer to observe temperatures in the various levels through which he falls.

Such a parachute sounding was made several years ago over the Caucasus Mountains in Southern Russia by an aerologist named Zechelin. He jumped from the plane at about 20,000 feet altitude in the late afternoon, when above two cloud layers (one around 8000 feet, the other around 15,000 feet). These cloud layers were caused by the warm front of a cyclone or 'low' that was centered nearby.:

An icy stream surrounded me as rose full height in the cabin the thermometer at — 12°C. [+11° F.]. I got out with sufficient ease — only a certain heaviness in the head and body indicated unusual conditions.

I stood as usual, one leg on the wing and the other on the fuselage step, took hold of parachute ring with right hand, and looked down. The ground was made almost invisible by a cloud layer which here and there showed big 'windows.'

Receiving a signal from the pilot, I jumped ...fell 150 meters (500 feet) before the parachute filled. As I was approaching the first layer of clouds, the swinging was not great; but afterwards it was entirely impossible to remain vertical. I was swinging . . . like a pendulum . . . at times bringing me horizontally up to the level of the parachute. From time to time I ceased falling altogether, held up by a rising current. There was a painful ringing in the ears, my face froze, and it was unpleasantly damp.

I entered the clouds; my goggles became fogged. When I dropped out of this cloud layer I rubbed my goggles, and saw the plane flying near! A few minutes later I entered the lower cloud layer. Turbulence and swinging were again very violent. I became tired and felt nauseated. Only the draft of cold air eased this seasick feeling. Dropping out of this cloud layer, I saw the ground below and prepared for a landing. Though the surface wind was quite strong, the jolt when I hit the ground was not very bad, and I was not injured.

Zechelin landed at a place about thirty miles from the point over which he tumbled out into space. During the nineteen-thirties, airplane sounding became an early-morning commonplace all over the United States. Under the Weather Bureau's expanded aerological programme, no less than twenty-four airplanes were making 'weather hops' daily from airports all over the country. About one third of these planes were furnished by the Army, one third by the Navy, and one third by civilians under contract with the Weather Bureau.

The planes usually climbed, weather permitting, to heights around sixteen or seventeen thousand feet. Usually dispensing with a separate observer, each plane had installed between the wings, in a cushioned aluminum box, a revolving drum aerograph (exactly like the kite aerograph in principle) that recorded values of pressure, temperature, and humidity against time.

The sounding itself took an hour or so, and another laborious hour was employed by ground aerologists to translate the aerograph curves into definite values of pressure, temperature, and humidity against altitude at all levels, in which final form the sounding could be used to plot charts and diagrams useful in weather analysis and forecasting. Thus these recorded airplane soundings were unnecessarily slow. Moreover, they were dangerous, and too expensive in planes and pilots. During one single year the Army, maintaining eight sounding stations, lost eight planes. For high-altitude soundings, the airplane has now been generally replaced by the radio sounding balloon or *raysonde*. But of these small radio robots.

Some of the airplane-sounding pilots were caught in terrible flying weather — dense, 'blind' cloud layers extending through miles of vertical height; terrific bumpiness and turbulence in thunderstorms; icing that formed on wings and propeller, loading the laboring plane until it staggered and fell out of control; high-level hurricanes that blew pilots in or above obscuring cloud layers all unknowing far off their assumed courses.

Many are the sagas of brave men fighting for their lives on the 'weather hop' while unknowing groundlings were sleeping soundly miles below. The getting up at 3 A.M. or so, the long, tiresome climb (particularly if flying blind) with insufficient oxygen, the tedious monotony of maintaining the climb indicator needle a 300 ft/min in a plane that could easily do 1000 ft/min or better — all these factors combined to make the pilot sleepy and unwatchful, and predisposed him towards accidents. One Navy sounding plane at Bolling Field (Anacostia), on landing one morning after the weather hop, turned over and burned up for no good reason whatever.

At stations along the east coast, weather flyers were always in danger of getting blown out to sea in unsuspected highaltitude gales. One of the Air Corps pilots at Boston, here called merely Lieutenant Smith, climbed one morning into a low overcast, heading west along the leg of the radio range that pointed towards Albany. The opposite leg of the 'on course' signal pointed out to sea, but Smith was not interested in that leg -or thought he wasn't. As he climbed into the overcast above the Bean City, his ground speed was very slow against a strong westerly gale. For many minutes he flew entirely blind through the extensive cloud layer, climbing steadily at the prescribed rate.

Gradually the Boston range signal sounded fainter and fainter — so Smith naturally assumed that he was making good his course to westward. When he reached three-mile altitude, still flying blind in 'pea soup,' the Boston range signal was very faint indeed. Smith, naturally assuming he was somewhere over central Massachusetts, cut his motor and dove steeply downward, heading eastward in what he thought was the direction of Boston. The altimeter needle unwound as he dropped rapidly to 8000, 6000, 4000. Then 2000, and still no sight of the ground. He was beginning to get anxious, for there are 1000-foot hills in central Massachusetts.

Then, suddenly, at 900 feet he broke through the bottom of the cloud layer. But beneath him was only an endless waste of gray and heaving water — the storm-tossed Atlantic. The next thing he saw was an ocean liner, apparently headed for Europe. He hurriedly turned the plane's nose westward. Here in the lower levels he could at least make some headway against the gale by opening up his throttle.

This he did, boring steadily westward at an air speed of one hundred and thirty and a ground speed of perhaps eighty miles an hour. After half an hour of anxious flying above the wave tops, he sighted the tip of Cape Cod. Another half-hour, and Smith landed at Boston Airport, forever convinced of the potency of the prevailing westerlies at high altitudes.

MEET THE SUNRISE

Radio balloons have now taken all the hazard out of the 'weather hop.' But they have also taken much of the virile adventure out of it, and some flying men will mourn the passing of a glamorous era in aviation. And in occasional crystal-clear weather, what marvelous sunrise views those flying weather men used to see. Here, for example, is the account of a weather hop I made as observer in December, 1934, from Anacostia Naval Air Station (old Boiling Field). By a turn of good fortune, that Navy pilot and I climbed into one of the clearest and most luminous sunrises that ever wakened the National Capital:

SUI-8872, a trim little biplane with fuselage of gleaming metal and wings of silvered fabric, used to be a respectable Navy scout before she got mixed up with this aerological business. Now, more often than not, her hangar is brightly lighted in the small hours. This black winter morning is no exception. Men are wheeling her out into the dark cold. They spin her inertia starter. Her motor sputters, coughs protestingly over its inhalation of chill air, and begins to rumble steadily, pouring blue flames out of the exhausts.

Her tail shivers a little under the cold blast from her prop. Orion and all the lesser constellations are bright in the black sky; the moon, nearly new, hangs high in the east, a ball of dim purple enclosed by a ring of bright gold. We are out on the dark field, surrounded by steady boundary lights. Now the floodlights are on, the motor roars, and '72 is racing across the frosted field that sparkles blue-white in the glare. She climbs smoothly, over the gleaming black river bordered with calm lights, over lamplit streets, banking easily the while in a wide turn.

As we climb, the horizon widens, suffused by the first, faint foreglow of dawn. Below, Washington is a million twinkling jewels set in geometrical patterns of integrated and exquisite beauty on a background of darkness. Airway course beacons are winking on the dark earth, out toward all the horizons. '72's blunt nose is pointed now north, now south, but always aspiringly upward, as the blue flames pour from her exhausts and the motor sings its steady song. And, as she steadily climbs, the dawn keeps flooding steadily out of the east — cold, white radiance over an horizon rimmed with faint rose, fading to pale blue above, where the moon is a ball of violet in its crescent of thin gold.

Gradually the radiance brightens as '72 ascends an invisible celestial slope, until her burnished struts are aglow with soft light. But the dim earth is still lost in darkness, black turning to dull gray. Unbelievably, it seems warmer. Strange and fascinating are the temperature variations in our great atmospheric ocean. People are sleeping this morning in freezing weather on the ground, while up here, seven thousand feet above them, it is springtime!

The dawn is ever brighter, the moon a thin crescent of silver above it, the distant ground a less-dark gray, on which leadcolored bays and rivers begin to appear. The lights of Washington now are a great irregular pattern of gold threads on gray wool. Up here, space is a dim, silent void waiting to take form, hesitant between chaos and creation. '72 is bathed in the pale radiance that trembles through this void, and her struts and wings gleam with reflected tints of blue-white and rose.

The unconcerned little aerograph box bounces around superciliously in its rubber-cord suspension between the wing struts. It doesn't see the splendid sunrise or the jeweled earth patterns — or care. It represents Technology. All it knows is: *pressure 800 mb., temperature 2° C., humidity 44 per cent,* and it keeps ruminating, in its dull little mechanical brain, with admirable concentration on these entities and nothing else, setting them down meticulously on the chart.

'72 comes from Technology, also, and was born, so to speak, in a wind tunnel. But in the gentle lift of her wings, the clean sweep of her fuselage, and the easy curves of her flight, there is also a world of Art. She is a machine whose heart beats out four hundred horsepower converted from burning gasoline; yet, with her delicacy and responsiveness, her graceful, sweeping progress, she is something of a woman — or a feminine bird.

Ten thousand feet, and I turn a valve below my seat, feeding oxygen to the pilot. (If the observer faints, it doesn't matter.) It is getting colder. Below, the varied ground patterns begin to stand out in harsher contrast, yellow and black and gray. The Potomac, turning the colour of new lead, winds down in looping curves from the western mountains, and its widening flow can be followed seventy miles southward to where it empties into Chesapeake Bay. The great bay itself, in full view from its northern head to its southern outlet, reveals all the innumerable points and coves along its western shore. Beyond the bay is the dark loom of land, and beyond that — but wait. Washington, now, is merely a gray relief map out of an atlas.

Up here at two miles and a half it is getting colder in earnest and hard to breathe in the bargain. Exertion of any kind, in this thinning air, is difficult. Even '72, vibrating with power, is climbing slowly and still more slowly. The domed shadow of the earth is sinking steadily in the western sky. The moon above has paled to thin silver. The east is a flood of bright orange, shot with radial white sun-rays from below the horizon, and golden plumes along the remote cirrus proclaim the Sun God's inexorable approach.

Fifteen thousand feet, and we meet the sunrise! Suddenly Sol is above the horizon, a great yellow blaze, a blinding disk too august for mortal eyes, pure radiance in which '72's bright metalwork sparkles and flames. (But, three dim miles below, the flat earth is still lost in gray shadow.) As we bank and turn in this rare new world of flooding sunlight, even the aerograph box gleams scarlet, catching a bit of poetry from the dawn. '72 fights for altitude. Her blunt nose is still pointed aspiringly upward toward a luminous sky of intense blue, but her path is almost horizontal, and the altimeter needle scarcely moves. It is bitter cold. We have sounded through springtime into harsh midwinter. Every breath is an effort.

Sixteen thousand feet! The earth and its waters are outspread at our feet. Chesapeake Bay resembles a narrow lake, barred across by the bright sun-path. Beyond it is a dark strip of unimpressive width, merely part of Maryland and all of Delaware. And beyond that, a hundred and twenty miles away, is the near shore of the Atlantic, whose restless waves are visibly afire in the blinding sun-path, far out to the haze-dimmed horizon!

The motor's dull roar dies to a soft rushing sound, as '72 points her nose thankfully downward toward the gray earth, just beginning to be touched by level sun-rays. Here and there on the flat surface ridiculous little trains are spinning long ropes of smoke. Thick ground fog rolls in billows along the curves of the river.

Buildings and monuments of beautiful Washington gleam white as alabaster in the sunrise. '72's nose points steeply downward as she dives for the wakening world below. Ten thousand feet, and time to turn off the pilot's oxygen. '72 is rolling happily from side to side now, like a hungry horse bound for the stables, as the wind whistles softly through her struts and wires. She knows she is headed for the comfortable ground and her warm hangar. She is dropping so fast that we must swallow hard and often to keep terrific pressure pains out of our ears. Now the pilot is throwing '72 around in steep, diving turns. Expanding, relief-map Washington appears first over her nose, then over one wing, then over the other. (If you don't keep swallowing, an eardrum may collapse.) We are still plummeting downward.

A last, wide turn, swooping over the fog-bound river, and we are landing on the misty field. Down from sixteen thousand feet in nine minutes: that is equivalent to falling steadily at twenty miles an hour. And something has happened to the sunrise. Surely, the sun seemed higher up aloft. Our descent has been so rapid that we have gone Joshua one better; the sun not only stood still, but went temporarily into reverse!

STABLE AND UNSTABLE AIR

Having now outlined just how soundings are made by means of kites, captive balloons, and climbing airplanes, it remains to examine how the data thus found — values of pressure, temperature, and humidity at various levels in the upper air — are employed by aerologists in their exacting but interesting business of analyzing and forecasting the weather.

Adiabatic Diagrams

Aerial soundings become most clear and meaningful when they are plotted in the form of charts or diagrams. The simplest and most direct of these charts, and for many purposes the most useful, is the adiabatic diagram, devised around 1910 by a German aerologist named Neuhoff. On the adiabatic diagram the sounding itself becomes a curved or irregular line that is not unlike, say, a curve showing the variations in business activity over a period of months.

On most familiar economic or humanistic charts, of course, the vertical ordinate is some human activity, such as business, and the horizontal ordinate is time. On the adiabatic chart the vertical ordinate is pressure (or what amounts to the same thing, altitude — since pressure always decreases with increasing altitude); and the horizontal ordinate is temperature. In addition, on a separate portion of the adiabatic chart, relative humidity may also be plotted as a horizontal ordinate.

The adiabatic diagram thus indicates clearly how temperature (and perhaps relative humidity also) varies with decreasing pressure or increasing altitude. The temperature *versus* pressure (altitude) curve shows at a glance all the often curious temperature changes in the various levels aloft, and where abnormalities occur — abnormalities such as temperature inversions, excessively steep lapse rates, and so on. On the adiabatic diagram, also, an aerologist can find critical points where there is a marked change in temperature, humidity, or other air-mass characteristics — a change that will perhaps indicate the boundary zone between two unlike air masses. Now, in order to see just how a sounding is plotted on an adiabatic diagram, suppose we recapitulate part of the military airplane sounding:

Altitude (ft.)	Temperature (°F.)	Lapse Rate (°F/1000 ft)	Relative Humidity (%)
20	79°		95
1000	74°	-5°	88
2000	76°	+2	85
3000	73°	-3°	82
4000	70°	-3°	80
5000	67°	-3°	78
6000	65°	-2°	76
7000	59°	-6°	81
8000	56°	-3°	80

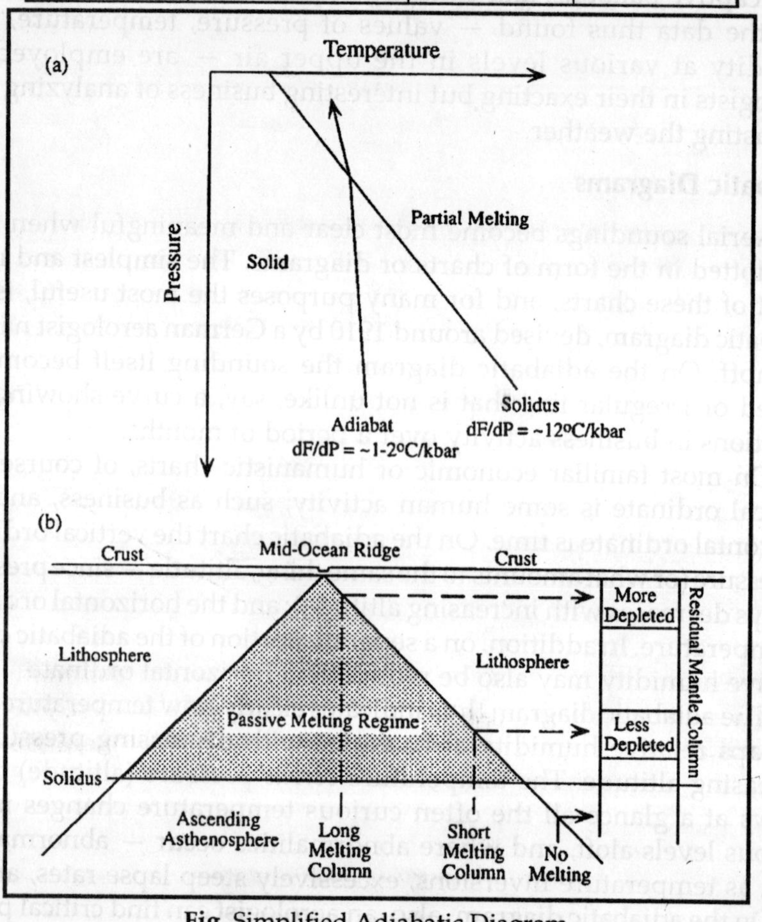

Fig. Simplified Adiabatic Diagram

In order to plot this sounding, suppose further that we rule off a diagram into checkerboard-like squares, marking each successively higher horizontal line 1000 feet, 2000 feet, and so on up to 8000 feet, and marking each vertical line from left to right 50°, 55°, and so on up to 80°. We then have a simplified adiabatic diagram on which this particular sounding can be plotted.

The steep temperature decrease in the first thousand feet of our sounding (lapse rate -5° F/1000 ft) now becomes a fortyfive-degree oblique slope of the temperature-altitude curve upward and to the left. The temperature inversion of the second thousand feet shows as a jog upward and to the right, like the profile of an overhanging cliff. The moderate temperature decrease of the third thousand feet shows as a steep slope upward and to the left.

An isothermal layer would appear as a vertical, cliff-like portion of the curve. Thus, paradoxically, the steeper the actual slope of the curve on the diagram, the more moderate is the lapse rate; and the steeper the lapse rate (or the faster temperature decreases with altitude), the more nearly horizontal is the slope of the curve. (This oppositeness of lapse rates and slopes is due, of course, to the conventional choice of altitude as the vertical ordinate, and temperature as the horizontal one.)

The relative humidity *versus* altitude curve is plotted at the right of the diagram in the same way as the temperature versus altitude curve. But the relative humidity units do not usually cover much space on the diagram, partly because the humidity curve is less important than the temperature curve, and partly because free-air humidity measurements are not very accurate anyway.

This description of how one particular sounding is plotted on an adiabatic diagram applies equally well to any other sounding. The horizontal lines are always labeled, of course, to accord with the altitudes (or pressures) actually sounded through, and the vertical lines are labeled to accord with the temperatures and humidities actually experienced. On a professional adiabatic diagram as drawn at a large forecasting station, the main horizontal lines might be labeled in hundred-millibar units of decreasing pressure, corresponding roughly to kilometers of increasing altitude; the main vertical temperature lines would be labeled for every ten degrees, Centigrade or absolute. Various other lines, useful in visualizing the changes brought about by rising air currents, in estimating where clouds will form, and so on, might also be printed on the diagram. But in principle this more complicated adiabatic chart would be no different from the simple diagram we have just examined.

Dry Adiabatic Cooling

With an understanding in principle of the adiabatic diagram and how soundings are plotted on it, we are ready to approach the slightly mathematical but tremendously important laws that govern the stability of an air layer. It might be easy and pleasant, in a book of this kind, to skip this whole matter of *stable and unstable air*. But if you skip it, you are missing the veriest fundamentals of the whole weather science. For stability, or lack of it, determines how likely the air layer in question is to overturn, or to encourage within itself various up-and-down drafts that may produce clouds, rain, and other varieties of weather.

As we have repeatedly seen, pressure in the air ocean varies with altitude — in the lower levels decreasing fairly uniformly at the rate of about one inch of mercury per thousand feet (100 millibars per kilometre). And on the average, temperature also decreases as one goes higher, by about 3° F. per thousand feet (about 6° C. per kilometre); although, as the actual soundings have abundantly shown, this average lapse rate seldom holds for any great distance in any actual air layer.

The easiest way to understand adiabatic cooling and stability or instability of air is to consider one small parcel or sample of air at a time — a sample of air which, experimentally and in our imagination, is made to rise or sink with respect to the air surrounding it.

If any sample of air ascends, the pressure around it decreases. Hence the air sample expands. Hence it cools. If this air sample loses no heat to and gains no heat from external things, it is said to cool 'adiabatically.' The *dry adiabatic cooling rate of ascending air* is always the same (provided no condensation of moisture occurs) — *about5° F. per thousand feet of ascent* (10° C. per kilometre). — (More exactly, the figure is 5½° F/1000 ft — but 5° is close enough for our purposes.) This fact (dry adiabatic cooling rate equals about 5° F/1000 ft) is one of the chief buildingblocks of physical aerology, and no intelligent aerologist, professional or amateur, can ever afford to forget it.

Dry-Air Stability

At any given pressure, cool air is relatively dense and heavy while warm air is relatively rare and light. Suppose that a certain layer of air has an existing actual lapse rate just equal to the adiabatic cooling rate (about -5° F/1000 ft). Any air sample that is made to rise from the bottom of this layer towards its top would of course cool by expansion at about 5° F. per 1000 feet of ascent.

Hence the rising sample would, actually, find itself always at the same temperature (and the same density) as the surrounding air. Hence it would have no pronounced tendency either to rise or to sink in the air layer. Any other air sample in the layer would show the same indifference to vertical motion. Hence the air layer as a whole is neither stable nor unstable — it is in a condition of *neutral stability*.

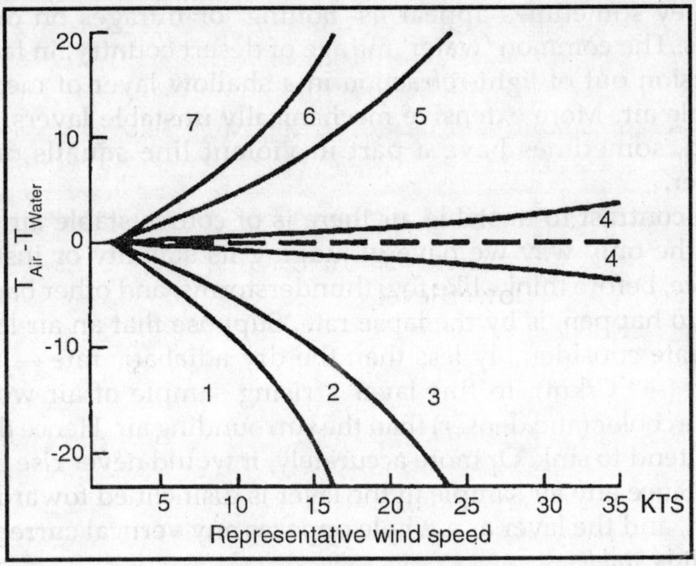

Fig. Stability of Air

Now suppose that an air layer has a lapse rate greater than the dry adiabatic rate — say around -7° F/1000 ft (-14° C/km). A sample of air rising through this layer, even though it cools steadily at 5° F. per thousand feet of ascent, is steadily becoming relatively less cool (and less dense) than the surrounding air.

Hence the air sample, constantly becoming lighter than its surroundings, would continue to rise of itself like a hot-air balloon, and the same is also true of any other air sample in the whole layer of air. Hence the entire layer is *unstable*. Because a layer having a lapse rate steeper than about -5° F/1000 ft is always unstable, whether any moisture is condensed or not, it is said to be *absolutely unstable*.

Moderately unstable air will not begin overturning on its own account, but if updrafts or downdrafts are once started or 'triggered off' by such commonplace causes as upward wind deflection over mountains, or intense surface heating by the sun, the overturning so triggered off will continue of its own accord and grow to a scale limited only by the depth and width of the unstable layer.

A very extreme case of absolute instability, called *mechanical instability*, arises from a lapse rate steeper than about -19° F/ 1000 ft (-34° C/km). In this case the upper air is actually denser to start with than the lower air, and the layer promptly overturns of its own accord like a topheavy boat. In the atmosphere mechanically unstable air layers of this sort are very infrequent, and those that do occur are usually of no great thickness. Nor can they exist for long.

They sometimes appear as 'boiling' or mirages on overheated deserts. The common 'water' mirage of desert country, in fact, creates its illusion out of light-refraction in a shallow layer of mechanically unstable air. More extensive mechanically unstable layers, very rare indeed, sometimes have a part in violent line squalls or tornado weather.

In contrast to unstable air there is of course stable air, and here again the only way we have of judging its stability or instability in advance, before things like fog, thunderstorms, and other bad weather begin to happen, is by the lapse rate. Suppose that an air layer has a lapse rate considerably less than the dry adiabatic rate — say -2° F/ 1000 ft (-4° C/km). In this layer a rising sample of air would soon become cooler (and denser) than the surrounding air. Hence the sample would tend to sink. Or more accurately, it would never rise in the first place. Since any air sample in the layer is disinclined towards vertical motion, and the layer as a whole opposes any vertical currents within itself, it is *stable.*

If the lapse rate is less than about -2½° F/1000 ft, the layer will remain stable under all conditions — regardless, that is, of whether any condensation of moisture takes place. Hence the layer is said to be *absolutely stable.* A marked but not uncommon case of absolute stability would be an 'isothermal layer,' in which temperature neither increases or decreases with increasing height, and the lapse rate is zero.

Still more stable, even, is a temperature inversion, in which temperature actually increases with height. Such a temperature inversion is so stable, in fact, that it usually acts as a pronounced barrier to all vertical convection — a sort of aerial 'lid' that stops ordinary rising currents, and forces rising smoke or growing clouds to spread out horizontally beneath the base of the inversion. But if the rising currents are violent, as in large cumulus clouds or thunderstorms, their very momentum may carry them through the inversion, which may be destroyed by their passage.

Moist Adiabatic Cooling

We have so far avoided of an air layer whose lapse rate is somewhere between -2½° F/ 1000 ft and -5° F/1000 ft. We have called it neither stable or unstable, for it can be either, depending upon the amount of moisture within it. This brings us to another fundamental building-block in physical aerology — *the moist adiabatic cooling rate,* which in lower levels (where temperatures are relatively high and moisture plentiful) is *about -2½°F/1000ft (-5°C/km).*

At higher and colder levels the moist adiabatic cooling rate may be as much as -3° F/1000 ft, or even -4° F/1000 ft. At very high and cold levels, indeed, it approaches the dry adiabatic cooling rate. It will bear repeating that the dry adiabatic cooling rate applies to air whose moisture remains in vaporous form; the moist adiabatic cooling rate applies to air whose moisture is condensing into liquid water (cloud droplets, raindrops, and so on).

Why should air in which moisture is condensing cool less rapidly than air whose moisture is remaining vaporous? Because of the 'latent heat' of condensing water. Each gram of water vapour, when it condenses to liquid water, releases nearly six hundred calories of heat. Where uncounted thousands of tons of water are being condensed, as they are in any sizable rainstorm, the actual amount of heat being released may be enormous.

Moist-Air Stability

It is now evident that, if an air layer has a lapse rate between -5° F/1000 ft and -2½° F/1000 ft, we cannot assess its stability or instability on the basis of temperature lapse rate alone. We must also consider its moisture — most conveniently, in the form of relative humidity at all levels. If relative humidity is very high, indicating the imminence of moisture condensation, even a lapse rate of -2½° F/1000 ft may be unstable — and of course any higher lapse rate would be even more unstable.

Hence a layer having an actual existing lapse rate between the dry adiabatic and moist adiabatic cooling rates (between -5° F/1000 ft and -2½° F/1000 ft in lower levels), is said to be *conditionally unstable.* If such a layer is nearly saturated with vaporous moisture, if its relative humidity is everywhere close to 100 per cent, then it is in effect unstable. On the other hand, if the layer is so dry that no condensation of moisture is possible, even with considerable lifting and cooling, then it is stable. Conditionally unstable air, in other words, behaves like stable air until condensation begins; after condensation begins, it behaves like unstable air. How conditional instability plays its part in the formation of clouds, rain, and stormy weather.

There is one other kind of instability that calls for passing mention here. But in the monstrous heavings of the air ocean, whole air layers, thousands of feet thick and weighing millions of tons, are often lifted bodily. If condensation of moisture occurs at the base of such a bodily lifted layer, but not at its top, the general effect of the lifting may be to steepen any pre-existing lapse rates within the layer. Hence a layer having a lapse rate steeper than -2½° F/1000 ft, perhaps conditionally unstable, may also or alternatively be *convectively unstable* — meaning that if the whole layer is lifted far enough by external forces (such as the undercutting wedge of colder, denser air that constitutes a 'cold front'), it may eventually become absolutely unstable. Ordinary instability is shown on an adiabatic diagram. Convective instability is shown on another type of diagram — the potential temperature-specific humidity diagram.

Before leaving this knotty but fundamental matter of stable and unstable air, it may be well to visualize the key facts in a diagram, and to summarize them in round numbers as one brief table:

Table. Stability of Air

Temperature Lapse Rate	Relative Humidity	Stability Behavior
Less than -2 ½° F/1000 ft	Immaterial	Absolutely stable
From -2 ½° to -5°	Low	Stable
From -2 ½° to -5°	High	Conditionally unstable; perhaps convectively unstable
Steeper than -5°	Immaterial	Absolutely unstable
Steeper than -19°	Immaterial	Mechanically unstable

Thus from a knowledge of the lapse rate, obtained by a sounding, we can determine stability, and perhaps predict what is likely to happen in some portion of the air ocean. But we can also work backward, from a knowledge of what is happening, to what the lapse rate must be. If clear-air winds are strong and gusty above broken country, air throughout the lower levels is being continually mixed by turbulence, and the lapse rate must be close to the dry adiabatic rate (about -5° F/1000 ft). When clouds are continuous and active through a certain layer, the lapse rate in that layer is necessarily something like -2½° F/1000 ft or -3° F/1000 ft. Towards the end of a calm clear night, a temperature inversion is certain to form in the lower levels as ground cooling goes on.

The simplest and most salient fundamentals of what is in professional practice a complex and varied technology. It is possible to squeeze far more meaning out of a sounding than we have indicated here by plotting it on various sorts of diagrams. One such is the potential temperature-specific humidity diagram already mentioned, whose full usefulness will appear in connection with air-mass-and-front analysis. It is also possible to plot many soundings together on more comprehensive charts. All these upper-air diagrams, of course, must be considered together with simultaneous weather maps.

RISING ROBOTS

The age of aerial exploration by man-carrying balloons or aerostats declined, smaller balloons were used to carry aloft recording instruments. These small aerial robots had soon ascended to heights far above those ever reached by men, and had discovered the existence of that great upper isothermal region known as the stratosphere. At first these sounding balloons or *ballon-sondes* were manufactured by cumbersome and expensive processes. But since about 1930, as related in the preceding and efficient sounding balloons of blown latex have been available for universal for upper-wind finding and sounding.

Sounding Balloons

Most present-day sounding balloons are of three uniform types or sizes. The smallest type is, in effect, a spherical bag of thin rubber, fitted with a strong neck for coupling to the inflation tube, about fourteen inches in (uninflated) diameter and weighing some three or four ounces. Inflated with hydrogen or helium to a free lift of about forty ounces and a (sea-level) diameter of about three and one half feet, this balloon will carry aloft a sounding instrument weighing up to ten ounces at a rise rate of more than one thousand feet per minute.

As the balloon rises it expands, of course, until it finally bursts, high in the stratosphere, at a diameter of seven or eight feet. Then the sounding instrument sinks quickly but safely back to earth by means of a small parachute that may be nothing more complicated than a three-foot square of thin cloth secured at the corners by four strings.

A larger type of balloon, suited to heavier instruments, weighs about a pound and, uninflated, is a rubber bag about two or three feet wide. Inflated to a starting diameter of six feet or so, it will carry aloft two pounds of 'pay load' at a good rise rate. A still larger balloon can be inflated to a starting diameter of eight feet or more, and will carry up apparatus as heavy as five pounds.

Ordinarily these balloons are expected to reach a height of ten miles or so before bursting. By under-inflating the balloon (and necessarily reducing its rise rate) it can be made to reach still higher levels. But the absolute height limit of this sounding method, imposed by the physical qualities of rubber, hydrogen, and air, is probably around twenty-five miles. In order to sound higher than this, men must at last realise in practice the immense theoretical possibilities of the high-altitude rocket.

Balloon Aerograph

Various types of balloon aerographs (for recording values of pressure, temperature, and humidity) have been devised. Most of the older ones include a clockwork-driven drum, and though of lighter construction, they are in principle similar to the kite and airplane aerographs we have already examined. But around 1930 a Belgian named Jaumotte invented what will probably remain for many years the cheapest, lightest, and most beautifully simple sounding aerograph in the world.

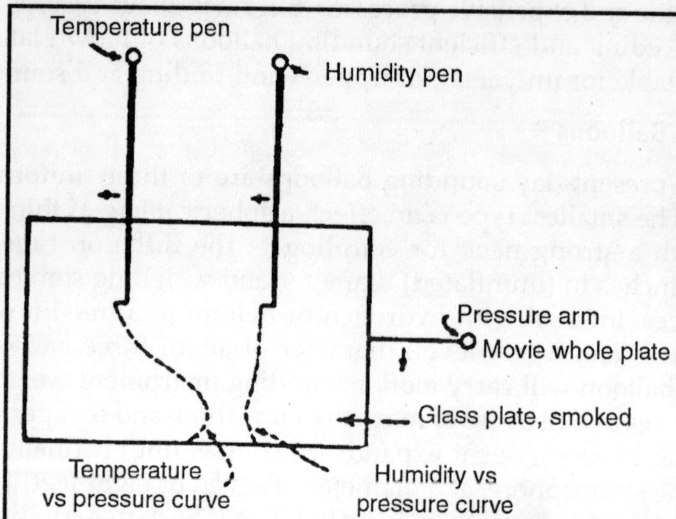

Fig. Elements of a Balloon Aerograph

No clumsy rotating drums for this man Jaumotte. Instead, as a recording surface for the instrument pens, he used a miniature glass plate only one by one and a half inches in size.

This little glass plate is smoked before the ascent. As the balloon rises, a small pressure element (similar to that in an aneroid barometer) moves the whole plate bodily in one direction. Meanwhile pens movable sideways and connected to the temperature and humidity elements scratch traces on the smoked glass in such a way that the result is, in effect, an adiabatic diagram in miniature. Upon recovery, it can be enlarged photographically to any convenient size for study. The entire aerograph is no bigger than a man's fist, and weighs less than two ounces complete with its aluminum-foil case.

The great difficulty about sounding balloons has always been getting them back — or more important, getting the data back — in time to be of practical use to weather men. The instrument usually carries a tag offering some few dollars' reward for its safe return; and in settled country, such as the American Midwest or eastern seaboard, perhaps four out of five released balloons are found by someone, somewhere, and actually returned. But the lapse of time from the ascent itself to the wanderer's return may be days or weeks — too long for the data to be considered in forecasting weather that has already come and gone. Ordinary robot-balloon sounding might be described by parodying an old sea song into: 'Sounding, sounding, into the aery main; and it will be full many a day ere you see that balloon again.' Hence aerologists have developed various methods and devices to get the data back immediately, regardless of what ultimately happens to the balloon.

Simplest of all are limited-height sounding balloons which burst or deflate at a predetermined height. Bursting may be accomplished by a small powder charge connected to a time fuse, or less spectacular deflation may result from a pressureoperated valve that releases the gas. If the device is set for twenty thousand feet altitude, say, the parachuting instrument will probably descend within telescope-sight of the ground station. Or if winds are high, the drifting balloon can perhaps be followed in an automobile or light airplane.

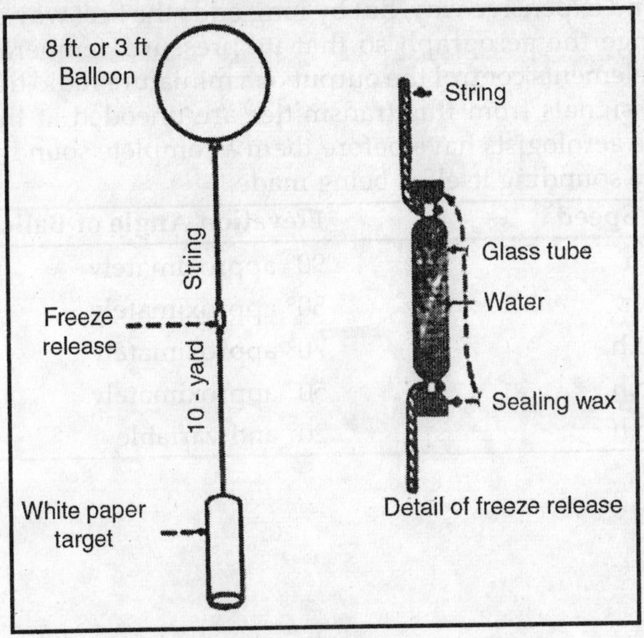

Fig. Balloon and Target for Finding the Freezing Level

Another ingenious gadget is the freeze-release. Connected in a string between balloon and paper target is either a metal tube bent into a circle or a straight glass tube, either being filled completely with water and sealed. As water expands by about ten per cent on freezing, the balloon's arrival at the freezing level will be signalized by the opening of the metal tube or the breaking of the glass tube. Either can be arranged to break the string connection and drop the target. As the balloon is being followed by a theodolite, the height of 32° F., or 0° C., is immediately known. Hence the observer knows also the average lapse rate up to that level — knowledge which may be useful in weather forecasting, aviation dispatching, or finding ballistic density for artillery.

Still another device, good for either night or day, is the thermoscope switch. A bi-metallic temperature element may be arranged, say, to close a switch momentarily for each ten-degree fall in the temperature. This electric contact flashes a light at night, or explodes a smoke-puff by day. The temperature element may also be arranged to close another switch if the temperature increases with height, flashing a light or firing a smokepuff of different colour. Thus a fair idea of the temperature distribution aloft is immediately transmitted to the observer.

There is still another way of flashing back the desired data -a more complex and expensive way, but by long odds the best way of all. This is to arrange the aerograph so that its pressure, temperature, and humidity elements control the output of a miniature radio transmitter. When the signals from this transmitter are uncoded at the ground station, the aerologists have before them a complete sounding record even as the sounding itself is being made.

Wind Speed	**Elevation Angle of Balloon**
0 m.p.h.	90° approximately
5 m.p.h.	80° approximately
10 m.p.h.	70° approximately
15 m.p.h.	50° approximately
20 m.p.h.	20° and variable

Chapter 10

Tropical Hurricanes

In the tropics marked fronts are rare. The air masses over these warm blue waters, bright green jungles, and burning deserts are mostly warm-moist (tropical maritime) or warm-dry (tropical continental or subsidence). As a result, the sun-drenched inhabitants enjoy most of the time a great sameness of tropical 'air-mass weather,' which inclines them towards unhurried living and 'mañana' habits of thought.

In winter (when the sun has left the zenith) — and carrying over well into the spring — there is the beautiful tropic *dry season* of clear skies and steady trade winds. In summer (as the broiling sun swings directly overhead in a brassy sky), and well into the fall, there is a season of steaming nights, of towering clouds and daily thundershowers, of northward-migrating doldrums marked by fickle breezes and strange calms — the tropic *rainy season*.

Here in the thundery doldrums, at intervals of a few days or a few weeks during the rainy season, there breed those dread, hundred-odd-mile-wide wind whirls known as tropical hurricanes or typhoons. For the moist, unstable air above these steaming areas of warm and azure sea is continually building up into towering showers and squalls marked by violent updrafts extending to high altitudes. If the rising currents happen to extend over a fairly large area at one time, and if the general aerological conditions are favorable, the latent heat released by large-scale condensation may cause a local drop in atmospheric pressure, which in turn induces a further inward flow of warm, moist air. Swayed by the earth's rotation, these inflowing currents soon assume a spiral or circular form.

The result is a destructive hurricane, driven by the gigantic energy of latent heat released in the wholesale condensation of water vapour; and before it eventually subsides in some far place, it may inspire no end of 'roaring deeps, and fiery sands and sinking ships, and praying hands.'

At the equator that fictitious aerological entity we call 'earthrotational-force' is zero, and hurricanes are impossible there. And north or south of about latitude 30° their initial formation is also impossible (due to lack of warm enough and moist enough air supply), though they sometimes wander as far as latitude 40° or even 50° before they are done with their violence. Hence most hurricanes form in latitudes between 10° and 25°, north or south — and most particularly along the great, world-encircling 'moisture front' that marks the boundary between the warm, relatively dry air of the trade wind belts and the also warm, ultra-moist air of the doldrums.

The initial impulse that spawns the meteorological monster along this trade wind — doldrum front may be the invasion of a large mass of modified polar air from higher latitudes, forming a sort of three-way junction between modified polar (cool-dry), tropical subsidence (warm-dry), and tropical maritime (warmmoist) air masses. Aerological conditions near such a threeway air-mass junction would at least favour the initial triggering off of a large air whirl. Whether they actually do start it, and if so just how, are questions to be fully answered only by means of many radio-sounding-balloon and radio-pilot-balloon ascents within and near actual forming hurricanes — a large practical assignment for any weather service, and a very expensive, difficult, and dangerous one.

Polar air outbreaks do, strange as it may seem, occasionally extend far into the deep tropics. I happened to observe such an outbreak from the *S.S. Timber Rush* in March, 1939, while in the Pacific around lat. 10° N., about a hundred miles south of El Salvador. Evening of March 6, before frontal passage, saw scattered clouds, with light variable winds and sea calm, temperature 81° F., dewpoint 74° F.

By midnight of the sixth-seventh a dry gale was blowing out of the northeast, kicking up a rough sea; the sky was crystal clear with myriads of southern stars twinkling in the high upper winds aloft; the temperature was down to 73° F. Next day the temperature was back to 79° F.; the dewpoint, however, was only 67° F., showing the passage of a 'moisture front.'

The geographic regions where hurricanes occur in the northern hemisphere are:

- The Caribbean Sea, Gulf of Mexico, and near-by seas and coasts ('West Indian' hurricanes);
- Off the West coast of Mexico;
- The western Pacific near the Philippines, China, and Japan (locally called 'typhoons');
- Off the east and west coasts of India (locally called 'cyclones').

In the southern hemisphere:

- East of South Africa;

- East of Australia;
- Northwest coast of Australia.

Hurricanes are curiously absent from the South Atlantic, probably because the doldrums never here extend far enough south of the equator. The name 'cyclone,' incidentally, originated in India from the likeness between the spiral winds of a hurricane and the coils of a snake. Some writers call hurricanes 'tropical cyclones,' and others ponderously call temperate cyclones 'extra-tropical cyclones.' But a better nomenclature is 'cyclone' to mean a large-scale, 500-mile-wide temperate-zone whirl, usually with fronts; and 'hurricane' (derived from a Carib Indian word meaning 'big wind') to denote the usually somewhat smaller (say 100-mile-wide) tropical whirl, always more nearly circular and generally frontless in its most active stages.

I. R. Tannehill of the United States Weather Bureau has made an extensive recent study of our West Indian hurricanes, and before him they were studied by another Weather Bureau man, I. M. Clive. Tannehill book, *Hurricanes*, includes one of the few existing full records of a hurricane birth out of a great womb of generally squally weather. This particular hurricane formed about one hundred and twenty miles northwest of the Panama Canal on October 14, 1926, and fortunately just happened to be monitored by well-placed ship reports for the next three days.

At first the nascent hurricane was merely a hundred-mile ellipse of more than usual cloudiness and scattered thunder squalls, with pressure lowered only five or so hundredths of an inch at the ill-defined centre. Next morning in the same place, the pressure had dropped another five hundredths (1½ mb), the cloudiness and squalliness were somewhat more pronounced, and the counter-clockwise winds around the circle were breezing up to force 3 to 4 (15-20 m.p.h.).

On the third morning the growing hurricane was still stationary or very slightly to northward; the central pressure was now 20/100 (7 mb) or so below normal, and the winds were up to force 5 (2-5 m.p.h.). On the fourth morning the hurricane, now smaller but more intense, had moved a hundred miles to northward; its central pressure was now 40/100 (14 mb) below normal and deepening rapidly; the circulating gales were up to force 6 to 8 (40 m.p.h.); the dense disk of dark nimbo-stratus cloud, pouring down torrential rain, surrounded the small, relatively clear and calm central 'eye.'

It was now a full-fledged meteorological monster that increased rapidly in pressure-depth and force, passing within the next few days northward over Havana and Bermuda with 100-m.p.h. winds and considerable damage.

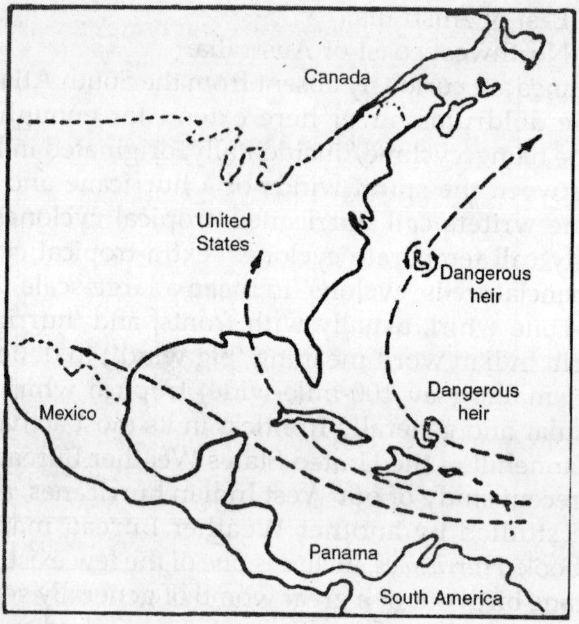

Fig. 54. Typical Eastern-American Hurricane Tracks

The hurricane birth described above is fairly typical of the species. Many a tropical storm, however, never reaches the full hurricane stage, spending its whole brief life as a small, rather mild cyclone with moderate circulating breezes. The average life of a hurricane is about ten days, though one that becomes a temperate cyclone along the general polar front may live for a month or so, traveling halfway around the world.

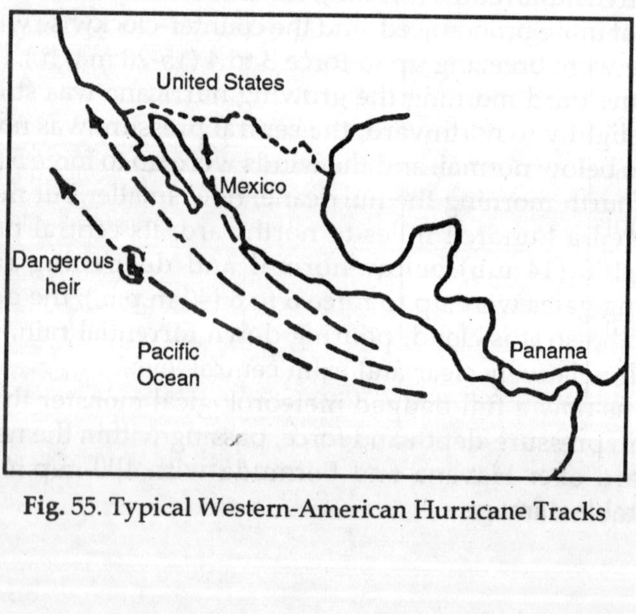

Fig. 55. Typical Western-American Hurricane Tracks

The usual tracks of eastern American hurricanes, like hurricanes elsewhere, are roughly parabolic in shape, curving constantly away from the equator. From its point of origin in the Caribbean or the western or middle tropical Atlantic, the low centre usually moves at first westward to northwestward in the belt between lat. 15° and lat. 20° N. The speed here may be anything from zero to twenty miles per hour, but is most likely to average about twelve miles per hour.

After reaching the horse latitudes (around lat. 30° N.), perhaps several days later, the usual West Indies whirl turns first gradually northward (slowing down somewhat in this recurve region), and then northeastward or eastward. Eventually it may move out over the Atlantic, ultimately becoming a temperate cyclone with fronts, and moving faster (up to 30 miles per hour or more), but decreasing in ferocity, as it reaches temperate latitudes. A fair proportion of West Indian whirls, however, never recurve eastward but run ashore northward or northwestward along the Gulf coast.

Western American hurricanes (somewhat smaller and feebler, by and large, than the Caribbean-Atlantic breed) usually move northwestward or north-northwestward up along the Mexican west coast at an average speed of around 10 miles per hour, keeping generally a couple of hundred miles or more offshore; they seldom live long enough to recurve completely or to penetrate far into temperate latitudes.

Aside from these more or less typical tracks, individual storms may describe very erratic courses, running ashore in unexpected places before or after the horse-latitudes recurve, describing loops, and so on. The general rule is that cyclones (including hurricanes) move away from or around or along high- pressure areas, and move towards low-pressure areas. Hence large highs to northward may block a hurricane's progress, or cause it to move straight west, or cast, or even southeast or south. A long low-pressure trough between two highs may be an attractive groove for an errant hurricane, as the astounded people of New England learned to their cost in 1938.

The energy that drives the screaming circular winds of a hurricane comes, as already indicated, almost entirely from the latent heat of condensing water vapour. Where millions of tons of aerial water are continuously condensing (as they are in a tropical cyclone), the awesome power of the released heat energy is equivalent of a conflagration that continuously burns millions of tons of coal.

If a hurricane runs ashore on the mainland it usually soon dies, because this colossal flow of latent-heat energy is cut off. True hurricanes thrive only where there is warm ocean immediately below

— tepid waters that supply warmth and moisture, and a flat surface that permits the easy spiral inflow of air in the lower levels.

THE SHAPE OF DESTRUCTION

The shape and motion of the whirling wind-and-cloud-andrain disk of a hurricane is somewhat like that of an ordinary phonograph record that spins rapidly and moves slowly sideways. For in the actual hurricane, likewise, the width of the whole whirl (say a hundred miles or more) is much greater than its thickness or depth (less than ten miles), and the speed of rotation (say one hundred miles per hour) is much greater than its speed of translation (say ten miles per hour). But the actual hurricane is not exactly a flat disk — it is bulged upward slightly at the centre, owing to the curvature of the earth's surface.

The size of a hurricane bears no direct relation to its intensity. Some very small hurricanes, twenty-five miles or so wide, have central pressures as low and hurricane winds as strong as much larger whirls, say, two hundred and fifty miles or more across. Upward the aerological effects of a hurricane (upper clouds and so on) extend to heights of six to eight miles, but the heavy rainfall and full hurricane winds are mostly within a mile of the surface.

The upper-air structure of a hurricane is, at the present time, surmised rather than known by adequate soundings. Through most of the storm, the air is rising fast as it whirls around, and at lower levels new air is coming in spirally from all sides to take its place. The ascending air blows generally outward again at cirrus levels (six to nine miles altitude) all around the storm, but some of it often blows out far ahead of the storm on the higher upper-air drifts. The general air drift at middle and upper levels of course determines the direction in which the whole storm moves. In the very centre of the great whirl there may be a core of descending air around ten miles wide (the calm central 'eye') which, heated adiabatically by compression, is somewhat warmer and drier than the encircling ring of rising and rotating air.

Hurricane Winds

While the roaring gales of a hurricane blow generally counterclockwise (in the northern hemisphere) about the low-pressure centre, there is of course considerable inward turning of the surface winds (about twenty degrees or so) due to friction with the ocean surface, so that the actual lower air flow is spirally inward and to some extent upward.

(The largest degree of inward turning occurs in the right rear sector.) But at gradient or lower cloud levels, the wind whirl is often truly circular. The heaviest crowding or convergence of air, and the strongest upward wind component (convection), occur in the right front sector; where also, as a result of this maximum convergence and convection, the heaviest rain usually falls.

If you are ever caught in a (northern hemisphere) hurricane, here is a general rule of thumb to locate the storm centre: *face directly into the wind; the centre will usually be on (or slightly behind) your right side.* This observed wind direction must be, of course, the average over a few minutes' time, exclusive of variable gusts. If you can see well enough to face the motion of lower clouds (the gradient wind) rather than the surface wind, the low centre should be directly to your right. (In the southern hemisphere, facing the wind would place the centre on, or slightly behind, your left side.)

What is the hurricane-wind picture as seen and felt by an observer on the ground or water, located directly in the storm's path? As the hurricane approaches, the characteristically gusty winds gradually increase. Meanwhile the pressure is falling more and more rapidly *while* the wind is rising — but the steep pressure fall does not (as many barometer-worshipers suppose) come much ahead of the wind. Hence the barometer seldom gives *positive* warning before the wind strikes.

As the storm centre approaches during several hours, the winds gradually increase to full hurricane force and may remain at this howling intensity for hours more. Then, very suddenly, the wind goes calm or nearly so for perhaps an hour or more, while a low, eerie rumbling sound around the horizon comes from the encircling wall of hurricane winds. *This calm central eye* has cruelly fooled thousands of tortured humans who were ignorant of weather fundamentals, and *has led many hundreds of people to their deaths.* For as the central eye passes, the screeching gale suddenly strikes again at equal or greater hurricane force, but this time from the opposite direction. Then, gradually, perhaps through many hours, the wind steadily decreases in force as the centre moves slowly on and the wearied observer comes at last to the 'hind' outer edge of the storm.

The force of the hurricane wind is of course everywhere proportional to the pressure gradient. The central hurricane pressure at sea level is usually below 29 inches (about 982 mb) -that is, more than 1 inch or 35 mb below normal. Central pressures below 27 inches (about 915 mb) — as much as 3 inches or 100 mb below normal — have been recorded.

The strongest hurricane winds so far observed (they cannot be accurately measured because the instruments blow away) have averaged around one hundred and fifty miles per hour for minutes on end, with occasional momentary gusts that appear to run as high as two hundred and fifty miles per hour! The steepest horizontal pressure gradients (averaging perhaps two hundredths of an inch per mile), along with the strongest winds, are usually fifteen to forty miles out from the actual centre.

The destructive effects of hurricane winds are terrible indeed, for at high speed 'thin air' becomes quite solid and tangible. In a severe hurricane exposed surfaces, such as house walls, may suffer wind pressures as high as a hundred pounds per square foot. In unsheltered locations the largest and stoutest trees are blown summarily down, and may be blown bodily away.

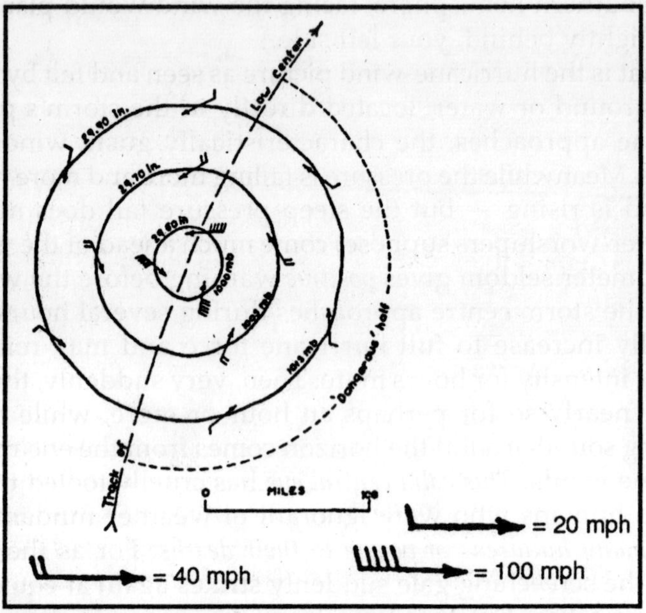

Fig. 56. Wind-and-Pressure System of a Typical Hurricane (Northern Hemisphere)

Ordinary frame houses are ripped piecemeal apart or crushed like eggshells. Modern steel-and-concrete buildings, however, can usually be counted on to hold fast, suffering nothing more than surface damage and blown-in windows. People and animals exposed in the open are usually soon smashed by flying débris. Even heavy stone houses (unless they are steel-reinforced concrete) may disintegrate under a 150-m.p.h. breeze.

Waves and Storm Tide

Along seacoasts, hurricane waves and tides may be even more

destructive than the 150-mile winds. The size of water waves bears a direct relation to the force of the wind that causes them. As a rough rule of thumb: time in seconds between crests of waves or swells equals force of originating wind in Beaufort units.

Thus ordinary storm swells, caused by force 6 to 7 (30-35 m.p.h.) winds, come at intervals of six to seven seconds. But hurricane swells come at slower intervals of ten to twelve seconds, indicating true hurricane winds of force 10 to 12 (more than 75 m.p.h.). These great swells, running out far ahead of the hurricane, are one of the most definite and positive warning signs of its approach at sea.

In addition to the battering of monstrous seas, there is in the centre and the right front sector of the hurricane whirl a marked general rise of water or 'storm tide,' occasioned partly by the lowered pressure, but mostly by the piling up of surface waters under the screeching wind. This piling up of water is most marked and destructive along a shelving coastline, and particularly violent along the left shore of a narrowing bay.

The actual rise of water may be ten feet or more above usual high tide, and storm tides of forty feet have been recorded. While the hurricane is distant this tidal rise is slow, but as the centre comes near, the waters rise more rap idly, and may roll over many square miles of ordinarily dry lowlands in a series of great 'tidal' waves that batter down buildings with driven wreckage, sweep ocean ships inland for miles, and destroy and drown all before them.

One of the worst hurricane tides ever experienced in the United States was the great Galveston 'flood' of September 8, 1900. Wrote I. M. Cline, who was in the thick of it:

The water rose at a steady rate from 3 P.M. until about 7:30 P.M., when [about an hour before the centre passed with a low barometer of 28.48 inches (965 mb)] there was a sudden rise of about 4 feet in as many seconds. The water at this time was about 8 inches deep in my residence and the sudden rise of 4 feet brought it above my waist before I could change my position. . . .

Before this sudden flood of salt water receded from the stricken city of Galveston, six thousand people were drowned.

Similar hurricane tides can occur around and over small islands in the open ocean. During the hurricane of January 13, 1903, the Society Islands (Tahiti and others) in the South Seas were swept by enormous waves that drowned a thousand Polynesians. Here, perhaps, was the germ of Nordhoff and Hall's best-selling South Seas novel, *The Hurricane*.

Perhaps the most terrible hurricane tide ever recorded swept the

Indian coast near the north end of the Bay of Bengal on October 7, 1737. Offshore the intense hurricane gales and seas sank twenty thousand ships of all sorts and sizes. Along the coast, the tide rose to forty feet above its usual high, and monstrous successive breaking waves, rising nightmare-like out of the ocean, swept and broke over the islands and low-lying mainland near the Hooghly River mouth. Three hundred thousand people perished.

Hurricane Clouds

The clouds around and near a hurricane centre are heavy nimbo-stratus and cumulo-nimbus extending up to several thousand feet altitude. Outside the heaviest rain areas, these nimbo-forms may shrink to strato-cumulus rolls pulled out along the gradient wind and driving rapidly with it. Farther aloft, and visible on the outer fringes of the storm, are alto-stratus or alto-cumulus, cirro-stratus or cirro-cumulus, and wispy citrus. The citrus, particularly, tends to stream out at high levels a couple of hundred miles ahead of the storm centre.

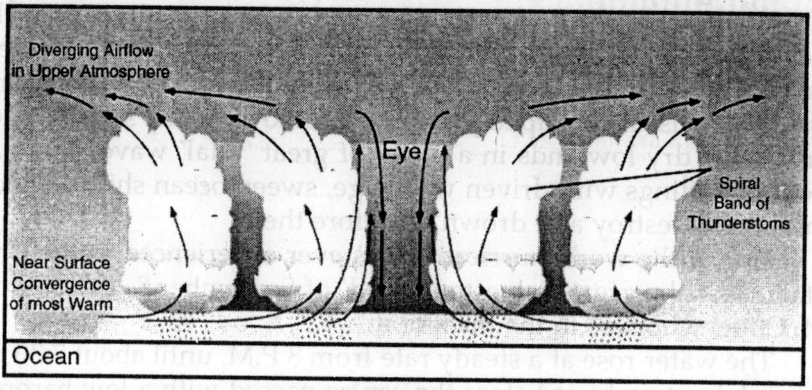

Fig. 58. Cloud System of a Typical Hurricane

Hence cirrus coming up over the horizon and seeming to radiate from a definite point is usually regarded in the tropics as a hurricane warning sign; though actually, you can often see plenty of citrus drifting about the tropical troposphere whether or not there is a hurricane in the vicinity.

To a more or less stationary observer on the ground or the water, at any rate, the first visible clouds of an approaching hurricane will likely be cirriform, and particularly a thin citrus haze that makes for fire-like, lurid, crimson sunrises and sunsets as well as solar and lunar halos. Then, after some hours or possibly two or three days, there may appear the 'bar of the hurricane' — an arc of dull, heavy gray cloud thrusting up over the horizon in the wake of the radiating citrus.

The cloud-bar is usually preceded, and may be entirely hidden, by detached but thickening fracto-stratus or 'scud' clouds in lower levels, driven rapidly by the gradient wind across the line of the hurricane's approach. If clearly seen the cloud-bar will show, by its motion along or steady growth above the horizon, how directly the hurricane centre is headed for the observer.

Torrents of Rain

Out of the condensation of unnumbered tons of water vapour sucked inward and upward by a hurricane there falls an unbelievable amount of rain. In the Puerto Rico hurricane of August, 1899, it was estimated that more than two and a half billion tons of liquid water fell on this hundred-mile-long island alone. In measured precipitation, that amounted to twenty inches or more at the rainiest locations, with nearly thirty inches reported at one place. Certain hurricanes of the South Indian Ocean have brought even heavier rainfall, with stations on the sea-girt island of Mauritius (lat. 20° S.) reporting more than twenty inches in twenty-four hours, and forty-seven inches within four days from one hurricane.

Thunderstorms and Tornadoes

It might seem as if the storm terrors already described were enough for any one no rural visitation. But hurricane showers are often also attended by blinding displays of lightning, though thunder scarcely makes itself heard above the roar of the winds. And rather rarely, a full- blown tornado or a waterspout is seen spinning merrily away and wreaking special havoc of its own in the midst of the wider destruction occasioned by its larger parent whirl.

SAILOR TAKE WARNING

Hurricanes are bad enough ashore, but it is on the ocean that one feels their full force and ferocity. Even a large steel steamship caught in a full-bodied hurricane is in for considerable trouble — passengers and crew members thrown about and injured by the wild motion, top-hamper and lifeboats swept away, hatches burst in, and so on. And if the waters are narrow, the largest ship may be swept willy-nilly ashore.

Smaller steamers and ocean-going windjammers may have to fight for their lives in the wild wind and wilder waves. Smaller craft, such as yachts or work boats less than a hundred feet long, have a very small chance of survival if a real hurricane catches them at sea. Even the staunchest sloop or yawl or ketch or schooner capsizes under bare poles in a 150-m.p.h. breeze, and even the sturdiest motor cruiser is rapidly swamped in the driving walls of wind-tortured water. Terrible as the wind ring about the centre is to shipping, the calm centre itself is considerably worse. For here in the chaotic central eye the seas (which are blown topless and held down by the hurricane winds around the circle) run at random every which way, out of all control or forbearance or reason.

Great waves coming from opposite directions may meet in one spouting geyser a hundred or more feet high. Monstrous pyramids of water rear themselves drunkenly up, and then topple senselessly down with force enough to burst in the hatches of any ship unlucky enough to be caught beneath.

So it is a fortunate thing that, owing to the researches of an American named W. C. Redfield in the eighteen-twenties, Colonel Reid in England, and many later investigators, the general wind-and-weather structure of hurricanes and the laws that govern their progress are now quite fully understood.

Dodging Destruction

The immediate precautions that a hurricane-harassed mariner must take, aside from battening down his vessel for a terrible beating, are three in number, as follows:

- Determine the bearing of the hurricane centre;
- Estimate its distance;
- Plot on the chart the ship's position, the hurricane center's position, and the probable path of the hurricane from a knowledge of their usual tracks in that region.

The smaller the boat, the harder all this will be to do, but the more a matter of life and death may be its accurate accomplishment.

Perhaps the quickest and easiest way to do this plotting, anywhere in the northern hemisphere, is by

- Copying the 'Windand-Pressure System of a Typical Hurricane' on a piece of tissue paper or sanded cellophane to the same scale as your chart,
- Moving the transparent hurricane diagram so drawn around on the chart until theoretical wind and barometer readings and changes (as indicated on the diagram) agree with the actual weather developments.

The problem can be simplified by eliminating the ship's motion from the calculations — that is, by heaving to for say three hours in order to get several successive observations from about the same place — and the time so lost may be well spent.

If radio weather reports can be exchanged with near-by ships, or if Weather Bureau broadcasts giving the position and expected movement of the storm centre can be copied, a much quicker plot of the hurricane center's position and path is of course possible.

The right half of the hurricane (as viewed by an observer looking forward along the hurricane track) *is called the 'dangerous half* — because, as the rotating winds are here added to the forward motion of the entire storm, actual gales in the dangerous half may run twenty to forty miles per hour higher than in other parts of the storm. Most of the rules for maneuvering ships at sea are concerned with keeping out of the centre and, if possible, away from the dangerous half. As a possibly useful rule of thumb for sensing which part of the hurricane is headed towards your position, the following may serve:

- If the wind increases but maintains its direction unchanged, you are probably directly on the track of the centre;
- If the wind direction (of origin) veers away from the first plotted bearing of the centre, you are likely getting into the safe half;
- If the wind direction veers towards the first plotted bearing of the centre, you are likely getting into the dangerous half. But better than any rule of thumb is knowledge of hurricane structure, careful observation, and graphical analysis.

A steamer or power-boat of any sort, being independent of ordinary winds, can at once proceed to put as much distance as possible between herself and the hurricane center's expected path (allowing for horse-lattitudes recurve and so on). If the vessel appears to be headed for the inner part of the dangerous half, this escape may involve a chancy crossing over into the safe half ahead of the hurricane centre.

A full-rigged sailing ship, dependent on the wind itself for her motion, has fewer and sterner choices to make. When the winds go above gale force she must usually heave to — 'on whichever tack permits the shifts of wind to draw aft' (so as to avoid being suddenly taken aback).

A small yacht or fishing boat, say forty to sixty feet long, would perhaps be best off drifting astern on a sea anchor, spilling out considerable oil if possible — or 'least badly off,' to put it more plainly — and a long and fervent spell of prayer to the Almighty might be fully as effective as anything further that could be done in the way of ordinary seamanship.

HURRICANE WARNING SERVICE

The United States Weather Bureau maintains, during the June-to-October season, a very complete and effective warning and forecast service on West Indian hurricanes. This service has been in effect, with gradual improvements, for a generation or more, having been started in the eighteen-seventies soon after the pioneering work of Father Viñes in Cuba. Special weather observations are now made during the hurricane season, morning and evening, at scores of land stations and on scores of ships scattered through the West Indies and over the Gulf of Mexico and the Caribbean Sea.

These reports are transmitted to the Weather Bureau Hurricane Forecast Centre by radio and wire, where they are digested and analyzed by a specially trained group of forecasters. Twice daily, or oftener, special bulletins giving the storm's position, intensity, and probable future course go out by radio to mariners and other interested people. All this applies to eastern American storms; the western American storms (off the Mexican coast) have so far been rather neglected.

Hurricane Portents

For various reasons (such as remote location, excessive static, or a defective radio), official hurricane warnings may not be available. Even if they are available, it is well to supplement them by observations of your own. Some of the portents that proclaim the approach of a hurricane — to isolated seamen, airmen, or landsmen alike. Suppose we now bring together in one place all this anticipatory weather-knowledge — knowledge that may, in a hurricane situation, save the life of him who heeds it. There are, first of all, preliminary wind signs and barometer signs that may 'cry wolf' falsely, yet may give the first observational inkling (in the absence of radio reports and broadcasts) of disaster to come. The wind sign, applicable anywhere in the trade-wind belts, is any marked change of direction or intensity in the regular trade, and particularly a falling off of the usually brisk trade into light, puffy airs.

The barometer sign, applicable anywhere within the tropics, is any marked interruption or change in the regular diurnal pressure tide (a normal total highlow twice-daily variation of about ten hundredths of an inch; maxima about 10 A.M. and 20 P.M.; minima about 4 A.M. and 4 P.M.). To note any significant barometer change and separate it from the diurnal tide, barometers in the tropics (and elsewhere) should be read mainly at 7 A.M. and P.M., and 1 A.M. and P.M.Now for the hurricane warning signs that may really mean the imminent approach of destruction:

- An unusually long and slow sea swell (ten seconds or more between crests) — observable, of course, only at sea or along the shore.

- A gradual abnormal rise of the average tide — observable only on the seacoast.
- Cirrus clouds radiating from a point below the horizon and constantly thickening and approaching from this direction.
- Brilliant crimson sunsets and sunrises, together with solar or lunar halos, perhaps combined with exceptional clearness of the lower atmosphere (exceptional visibility at sea, and so on).
- Change of normal wind or calm to gusty and fitful and increasing winds.
- Rapidly' increasing fall of the barometer — say three or four hundredths of an inch (1 mb) per hour, then six or seven hundredths (2 mb) per hour, and so on — to values below 29 inches (about 980 mb); together with steadily increasing and characteristically gusty winds.
- A heavy pall of dark cloud (the 'bar' of the hurricane) lifting above the horizon and approaching, usually preceded by scattered clouds, squalls, and showers.

CARIBBEAN AND GULF

A glance at any chart of eastern American hurricane tracks will reveal that the Caribbean, the Gulf, and the near-by Sargasso region of the Atlantic — our own warm, blue 'south seas' — are their natural playground. Through the middle of this hurricane region runs the great island chain of the West Indies; the shores of Mexico and the southern United States bound it to westward and northward. Hence practically all the islands and coasts in this region have been devastated at one time or another, and every year find hurricanes active here.

In 1932 an unseasonably late hurricane caught the British freighter *Phemius*, an able steel steamship of several thousand tons displacement, south of Cuba on November 5. The ship was disabled and carried for many hours along with the worst of the storm.

A fierce hurricane was blowing and a very high sea running. The ship was enveloped in spindrift, reducing the visibility to Nil, the No. 1 hatch not being visible from the bridge. The vessel was rolling heavily, the helm being of little use. So great was the force of the wind that shortly before 8 P.M. the funnel was blown overboard.

The ship was rendered helpless and from then on was carried with the hurricane in an unmanageable state. It would not be overestimating to put the wind force at 200 miles per hour. Hatches were blown overboard like matchwood, derricks and lifeboats wrecked, upper and lower bridges. Some of these West Indian hurricanes, failing to recurve at all, go ashore on the Mexican coast. In late August, 1938, for example, a storm centre passed over Progreso on the Yucatan Peninsula, bringing a quarter hour of calm between opposing gales.

Still other hurricanes go ashore on the Gulf coast of the United States. The first hurricane of 1939, in mid-June, passed northward over Cozumel Island (off Yucatan) on the twelfth. Next day it rocked the American steamship *Alabama* in lat. 25° N. long. 85° W. with barely-hurricane winds of force 8-9. Curiously enough, the storm was even then on the way to the state of Alabama, though it dawdled around the Gulf, and even described a complete loop, before getting there. On the sixteenth, now a temperate cyclone with fronts, it passed inland at Mobile and continued with diminishing force north-northwestward across Mississippi, where it merged with another temperate cyclone.

Still other hurricanes recurve early and pass northeastward far out in the Atlantic. In late October, 1926, such a storm, appearing out of nowhere, struck the Bermuda Islands with practically no previous warning; the previous weather had been marked by light variable winds, a steady barometer, and some high cloudiness.

Then the wind breezed up from the southwest with increasing cloudiness and rain — all of which looked like an ordinary temperate cyclone passing to northward. But next morning the wind backed to east-northeast with heavy increasing gusts as the barometer tumbled. Towards noon the central eye passed with light winds. After this comparative calm, the full hurricane blast smote at one hundred and twenty miles per hour out of the northwest, damaging more than a third of the houses on the island.

EAST COAST AND WEST COAST

The eastern American hurricanes that bring the shape of destruction before the largest numbers of people, and cause the greatest amount of mass-human consternation, are those occasional wind whirls which move at full intensity up along the Atlantic coast of the United States.

An off-course hurricane that moved up the cast coast in August, 1936, happened to be observed by D. W. Tomlinson from a high-altitude research airplane of Transcontinental and Western Air, Inc., en route from Kansas City to Newark. Viewed from the 'overweather' plane at five miles altitude, the storm had the shape of a great umbrella 35,000, feet high — an upper canopy of citrus and cirro-stratus cloud spreading out in all directions over a lower and denser canopy of rain cloud:

Thin and tenuous at the outer fringe, this canopy of citrus [cirro-stratus and alto-stratus] cloud thickened [towards the centre] until it joined the turbulent lower layers at 25,000 feet in a vicious looking melee of angry air which appeared black, purple and almost green in places. From this central maelstrom a lower layer spread out in all directions, tapering downward from 25,000 feet at the centre [over southern New Jersey] to about 15,000 feet over central Pennsylvania.

By far the most memorable cast-coast hurricane of recent history was the freak tropical storm which, on September 21, 1938, left the usual Atlantic track and raced north-northwest across Long Island and New England.

The cause of its fatal aberration was a long, north-south low-pressure trough between two large highs, marking the rapid northward flow of a wide, deep current of warm, unstable tropical air aloft. Brisk cyclones of tropical origin cross New England nearly every year, attended by heavy rain and whistling gales, but full-bodied hurricanes come less than once in a century. The last one had struck in September, 1815, the one before that in August 1635. People had forgotten that a hurricane could, in a rare coincidence of aerological conditions, visit this densely populated region.

This hurricane had been followed by weather services most of the way across the Atlantic from the Cape Verde Islands. It had moved steadily westward along the twenty-first parallel, curving northwestward well outside the Bahamas. It started the usual horse-latitudes recurve (which should have carried it well out into the Atlantic again) in lat. 36° N. during the evening of September 20. Its speed had been consistently above normal (fifteen to twenty miles per hour) – which circumstance, as well as the unusual aerological conditions aloft, had attracted the anxious attention of several eastern forecasters.

During the night of the twentieth to twenty-first the hurricane joined with an existing north-south front along the pressure trough and technically became a temperate cyclone. But owing to the peculiarly unstable distribution of air masses in its path, the wind whirl was maintained at full hurricane intensity and, unlike most hurricanes, was to lose little of its destructive force for some time after running ashore. In the early morning of the twenty-first the hurricane centre was off Cape Hatteras, veering northward again from its abandoned recurve, and beginning to speed up beyond all precedent.

By noon the hurricane centre was about a hundred miles off the Maryland-Delaware shore, and easterly-to-northerly gales were sweeping the whole coast from Cape Cod to Cape Hatteras. During the next three hours the centre raced due north along the seventy-third meridian at seventy miles an hour, crossing Long Island at 3 P.M. and passing inland over New Haven within half an hour. That afternoon and evening it sawed a broad band of destruction up through the backbone of New England. It crossed into Canada around midnight, and finally died out in the Quebec wilderness next day.

As it struck Long Island the central pressure of this storm averaged 28 inches (948 mb), and the hurricane winds ran well over one hundred and twenty miles per hour. The calm central eye, which had been evident over the ocean, was soon obliterated by surface friction over the land.

Most of the devastation naturally occurred in the right or 'dangerous' half of this fast-moving storm, along a zone extending for a hundred miles or so cast of the center's path. At crowded summer resorts along the south shore of Long Island scores of people were surprised and drowned as the waters rose ten to fifteen feet above normal high tide and hurricane breakers thirty to forty feet high swept across the sand dunes between ocean and bay.

As the storm moved inland similar destruction swept the coasts of eastern Connecticut and Rhode Island. People clinging to the highest dunes were swept away, along with great sectors of the sandy shoreline itself, by the driving seas. Seagoing ships were washed across railroads and highways. Providence was flooded with salt water, blacked out by electric failures, shaken by the explosion of a great gas-storage tank. Torrential rains and local fresh-water floods swept all of southern New England and New York.

On the 12.000-foot Empire State Building in New York City, seventy-five miles west of the hurricane centre in the 'safe" half, shrieking northerly gusts registered 120 m.p.h. as the world's greatest skyscraper swayed four inches out from vertical under the unaccustomed pressure (eight more inches of sway, and it would have crashed to the ground). At Harvard's observatory On 500-foot Blue Hill near Boston, in the 'dangerous half' a hundred miles cast of the centre, the southeasterly wind roared at a measured 120 m.p.h. for minutes on end, with occasional gusts over 150 m.p.h.

All across that broad populous region steel towers and radio masts were laid flat, thousands of houses were blown bodily about or crushed like eggshells, and millions of trees were blown down and scattered around like straw. Far up into Vermont and New Hampshire the appalling tree-blowdowns continued, lying in scattered swaths wherever the shape of the terrain served to intensify the roaring winds.

In this New England hurricane of 1938 about six hundred people were drowned by waves or killed by winds — a toll that might have been far larger in the absence of alert and heroic rescue services. As to property damage, it was the costliest disaster in all United States history with a total loss of four hundred million dollars — including twenty-six thousand automobiles, twenty thousand miles of electric lines, and two hundred and seventy-five million trees!

Less well known than the tropical storms of our southern waters and southern and eastern coasts are Pacific Ocean hurricanes off the west coast of Mexico. The west-coast storms are, by and large, somewhat smaller and less formidable than Caribbean-Gulf-Atlantic whirls; and, moving generally along an uninhabited sea track somewhat outside the usual shipping routes, they come less frequently under the detailed observation of mariners and weather men. Nevertheless they are true tropical hurricanes that thrive in the same season as their eastern American relatives, and fully developed specimens can equal the concentrated ferocity, if not the size, of any hurricane or typhoon.

Most typical west-coast hurricanes originate in or near the outer Gulf of Tehuantepec. Thence they move generally northwestward at something like ten miles per hour, gradually curving towards the north-northwest or north. After six or eight days, a typical storm dies out in the outer Gulf of California or in the Pacific off Lower California. Hence the most usual track is generally parallel to the Mexican coast, and two hundred miles or more offshore. Individual storms can, of course, depart widely from this general pattern.

Sometimes they curve to the northward or northeastward, running ashore along the Mexican coast and quickly subsiding among the high coastal mountains. On a rare occasion one of these storms continues northward and strikes the coast of Southern California. Other Pacific hurricanes may originate and move five hundred miles or so off the Mexican coast, and these storms may continue northwestward well out into the Pacific.

In the British freighter *Lochmonar*, I was able to study a more typical west-coast whirl from its outer fringes for several successive days of early September, 1939. We were bound from Panama to Los Angeles (fifty miles or so seaward from the usual offshore steamer track on account of possible enemy submarines), and first came into the hurricane area on the long west-northwest reach past Cape Corrientes. Then, turning northwestward near the usual 'corner' in these waters, we steamed parallel to the hurricane center's track at a speed slightly greater than its own, remaining in the innocuous outer fringes of its 'dangerous' half and drawing steadily ahead of it at a relative speed of fifty to a hundred miles a day.

Off Cape Corrientes on the morning of September 2, the wind was light easterly and southeasterly and the clouds high and scattered to broken. The only hurricane sign was a long swell (ten seconds between crests) rolling up from the southwest. At noon we received radio weather reports from several other ships along the coastal route; two of these ships, located one and two hundred miles up ahead, reported overcast weather with rising easterly winds and heavy southerly swells.

During the afternoon, as we approached somewhat nearer to the centre, the clouds thickened out of the west into a solid upper sheet of altostratus, with broken cumulus below; the wind swung more southeasterly and freshened. Putting all this evidence together, it appeared that the low centre was likely located two or three hundred miles west of us; and that on converging routes we would close somewhat with it before morning, though probably not enough for serious concern or a change in course.

During the night the clouds thickened and lowered, and a light rain began. By morning it was raining hard and steadily, the wind was still moderate southeast, and the long swell now rolled up out of the south on our port quarter. The barometer had fallen slightly, to 29.80 inches. These signs, and reports from other ships, placed the centre about two hundred miles away on our port beam to southwestward.

As we plowed steadily northwestward, aided by the long following combers, the low clouds and rain continued. But the wind drew gradually back into the east and the swell came now from cast of south, showing that the centre was dropping astern and turning somewhat inshore across the course we had already run. From then on the clouds and rain gradually lightened as we drew steadily ahead of the centre.

NIGHTMARE OVER FLORIDA

One state of the United States extends far enough into the West Indian hurricane belt to accept tropical storms almost as a matter of course — the great peninsula of Florida. Here, as in the West Indies, people are more or less ready for the worst.

One of the worst Florida hurricanes in recent years was the storm of early September, 1935. Appearing out of the Sargasso Sea east of the Bahamas, it moved northwestward across the Florida Keys, at the southern tip of the peninsula, during the evening of September 2. It was a small storm, but an extraordinarily violent one — one of the most violent ever recorded.

At the Florida Keys the central pressure was less than 27 inches (915 mb), and the circular winds were one hundred and fifty to two hundred miles an hour with gusts probably well over two hundred. The wind damage was almost tornadic in degree. Recurving in the Gulf, the storm cut northeastward across Georgia and the Carolinas and moved on across the Atlantic to Greenland as a larger but less intense temperate cyclone.

The whirl caught three large steamships in Florida Straits and treated them like peanut shells. A big motor-ship was totally disabled by wind and sea, stripped and partly flooded, carried willy-nilly across Alligator Reef and on shoreward for nearly four miles. A large passenger liner was driven helplessly aground on French Reef. The tanker Pueblo was disabled and blown completely around the storm centre in the space of eight hours. By good luck and good seamanship the people in these disabled vessels were saved.

Tragedy struck on the low-lying Keys, where three veterans' work camps housed seven hundred-odd men, to say nothing of many civilians who lived at near-by fishing camps. In accordance with Weather Bureau warnings, a rescue train started south on September 2 over the hundred-mile inter-key causeway of the East Coast Railway. The plan was to evacuate the veterans and civilians from the most exposed keys. But the train was delayed, reaching the camps only an hour before the storm centre.

It was swept from the track and destroyed by wind and water, along with the camps; and the concrete causeway itself, twenty feet above high water, was everywhere breached and broken down. At Long Key and Matecumbe Key the storm tide rose fifteen or twenty feet above the low-water mark, and the hurricane breakers rolled in like tidal waves. More than four hundred people perished in the veterans' camps and elsewhere on the Keys.

Another violent Florida hurricane — and one of the most disastrous storms on record — drove squarely across the thickly settled area around Miami on September 18, 1926. Like the Florida Keys storm just described, it struck during the dark hours (though the greatest wind force came after sunrise). The lowest pressure at Miami, within the central eye, was 27.61 inches (about 930 mb), and some of the recorded winds were nearly one hundred and forty miles an hour. The staggering toll was two hundred and forty-two people dead, perhaps three thousand injured, and a hundred million dollars in damage.

As usual on the Florida east coast, the hurricane came in from the Atlantic, approaching from the east-southeast. During the afternoon of September 17 the growing storm had passed close to Nassau in the Bahamas, blowing coconuts about like rain and bombarding yachts in the harbor with these storm-sped cannon balls. Weather Bureau warning of a northeast storm, based on fragmentary ship and island weather reports to eastward, had been out since noon.

Yet there were few signs, before evening, that a real hurricane was approaching Miami. The barometer had been falling steadily, but not alarmingly, all afternoon; as for the wind, it was less than twenty miles per hour at sunset. The small but immensely violent whirl approaching from seaward would strike suddenly, devastating the whole sixty-mile coastal region between Miami and Palm Beach.

By 10 P.M. the barometer began falling in earnest, and after midnight tumbled at the phenomenal rate of more than a quarter inch per hour. Meanwhile the northeast wind steadily increased, telephone lines went out, and general damage and destruction began.

The storm tide rose and waves washed entirely over Miami Beach to a depth of several feet, battering in doors and driving people to the upper floors of the half-flooded steel-frame hotels, where blown-in windows spattered them like shrapnel. Along the Miami shore of Biscayne Bay scores of vachts, tugs and other boats, including a United States Navy destroyer, were driven across the boulevard and many yards inshore.

Then, at about sunrise, came the treacherous calm centre. The shrieking winds lulled or ceased altogether. The sky brightened. Many people who had huddled in steel-frame downtown office buildings through the night rushed out into the streets to see the damage and rescue the injured. Thousands dug themselves out of the wreckage of their homes.

Hundreds, believing the storm over, started for work. Scores of pleasure-seekers crossed the long causeway out to Miami Beach for an early morning swim in the high surf. But all these people, as the Weather Bureau man was telling those nearest him, were playing a long chance on their ignorance of hurricane structure. This weather-ignorance was to cost many of them, and most of all the ill-advised excursionists to Miami Beach, their lives.

About 7.30 A.M., after an hour or so of lull, the other side of the encircling wind-wall (erroneously mentioned as a 'second storm' or 'second hurricane' in the newspaper accounts) struck out of the south. Again the sea rose over Miami Beach, where several hundred people were romping and frolicking on the sands, and many of the frolickers were swept away and drowned.

About a hundred erstwhile bathers were well out on the long causeway, headed back towards the city, when a great wall of water rose out of Biscayne Bay and washed over their heads, carrying most of them to destruction. In the city again, as during the small hours, 'limbs and trunks of trees, roofs of houses, and great big timbers were sailing along through the air; while telephone poles and palm trees would snap off close to the ground.' The steel-and-concrete building of the *Miami Tribune* was twisted so that it hung over the street. The braced steel beams of a building under construction were buckled badly out of shape.

Through that rueful day the southerly winds gradually diminished as the storm centre continued northwestward, spilling half of Lake Okeechobee over its lee shores and drowning many people there. Two days later, growing weaker, it could only muster eighty- to a hundred-mile gales at Pensacola and Mobile, and finally subsided over the mainland.

Chapter 11

Volcanoes

Volcano, an opening in the earth's crust, through which heated matter is brought, permanently or temporarily, from the interior of the earth to the surface, where it usually forms a hill, more or less conical in shape, and generally with a hollow or crater at the top. This hill, though not an essential part of the volcanic mechanism, is what is commonly called the volcano.

The name seems to have been applied originally to Etna and some of the Lipari Islands, which were regarded as the seats of Hephaestus, a Greek divinity identified with Vulcan, the god of fire in Roman mythology. All the phenomena connected directly or indirectly with volcanic activity are comprised under the general designation of *vulcanism* or *vulcanicity - words* which are also written less familiarly as volcanism and volcanicity; whilst the study of the phenomena forms a department of natural knowledge known as *vulcanology*. Vulcanicity is the chief superficial expression of the earth's internal igneous activity.

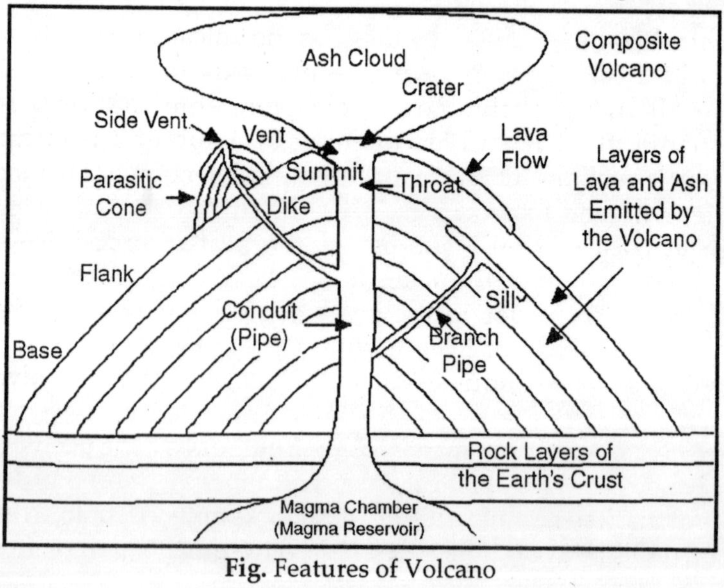

Fig. Features of Volcano

It may happen that a volcano will remain for a long period in a state of moderate though variable activity, as illustrated by the normal condition of Stromboli, one of the Lipari Islands; but in most volcanoes the activity is more decidedly intermittent, paroxysms of greater or less violence occurring after intervals of comparative, or even complete, repose.

If the period of quiescence has been very protracted, the renewed activity is apt to be exceptionally violent. Thus, Krakatoa before the great eruption of 1883 had been dormant for something like two centuries, and it is believed that the Japanese volcano Bandaisan previously to the gigantic outburst of 1888 had been silent for more than a thousand years.

A volcano may indeed remain so long dormant as to be mistaken for one completely extinct. The volcanoes of central France are regarded as extinct, inasmuch as no authentic historical record of any eruption is known, but there are not wanting signs that in some parts of this volcanic region the subterranean forces may yet be slumbering rather than dead.

PREMONITORY SYMPTOMS

A volcanic eruption is usually preceded by certain symptoms, of which the most common are local earthquakes. The mountain, or other eruptive centre, may be thrown by internal activity into a state of tremor; the tremors perhaps continuing intermittently for months or even years, and becoming more frequent and violent as the crisis approaches.

At first they are usually confined to the volcano and its immediate neighbourhood, but may subsequently extend to a considerable distance, though probably never developing into earthquakes of the first magnitude. The sudden opening of a subterranean crack, by rupture of a rock under strain, or the rapid injection of lava into such a fissure, will tend to produce a jar at the surface.

For at least sixteen years before the first recorded eruption of Vesuvius in A.D. 79 earthquakes had been frequent in the Campania and had wrought havoc in the cities of Herculaneum and Pompeii. Again, the formation of Monte Nuovo, near Pozzuoli, in 1538, was heralded by local earthquakes beginning several years in advance of the eruption.

So too in recent years many volcanic outbursts have been preceded by a succession of earthquakes; but as volcanoes are frequently situated in areas of marked seismic activity, the shocks antecedent to an eruption may not, unless exceptionally violent, receive much attention from local observers. It commonly happens that a volcanic outburst is announced by subterranean roaring and rumbling, often compared to thunder or the discharge of artillery underground. Other precursory symptoms may be afforded by neighbouring springs, which not unusually flow with diminished volume, or even fail altogether. Possibly fissures open underground and drain off the water from the springs and wells in the immediate locality.

Occasionally, however, an increased flow has been recorded. In some cases thermal springs make their appearance, whilst the temperature of any existing warm springs may be increased, and perhaps carbon dioxide be evolved. A disturbed state of the atmosphere is by no means a constant forerunner of an eruption, some of the greatest outbursts having occurred in a period of atmospheric stability: indeed the air is often felt to be close and still.

Immediately before a renewed outburst in an old volcano, the floor of the crater is generally upheaved to a greater or less extent, whilst the discharge of vapour from any fumaroles is increased. Where a crater has been occupied by water, forming a crater-lake, the water on the approach of an eruption becomes warm, evolves visible vapour, and may even boil. In the case of cones which are capped with snow, the internal heat of the rising lava usually causes a rapid melting of the snow-cap, resulting perhaps in a disastrous deluge.

It seems probable that by attention to the premonitory symptoms a careful local observer might in many cases foretell an eruption.

It generally happens that a great eruption is preceded by a preliminary phase of feeble activity. Thus, the gigantic catastrophe at Krakatoa on the 27th of August 1883, so far from having been a sudden outburst, was the culmination of a state of excitement, sometimes moderate and sometimes violent, which had been in progress for several months.

Emission of Vapour

Of all volcanic phenomena the most constant is the emission of vapour. It is one of the earliest features of an eruption; it persists during the paroxysms, attaining often to prodigious volume; and it lingers as the last relic of an outburst, so that long after the ejection of ashes and lava has ceased an occasional puff of vapour may be the only memento of the disturbance.

By far the greatest proportion of the vapour is steam, which sometimes occurs almost to the exclusion of other gaseous products. Such, at least, is the usual and probably correct view, though it is opposed by A. Brun, who regards the volcanic vapours as chiefly composed of chlorides with steam in only subordinate amount. In the case of a mild eruption, like that occurring normally at Stromboli, the vapours may be discharged in periodical puffs, marking the explosion of bubbles rising more or less rhythmically from the seething lava in the volcanic cauldron. S. Wise observed at the volcano of Sangay, in Ecuador, no fewer than 267 explosions in the course of an hour, the vapour here being associated, as is so often the case, with ashes.

During a violent eruption the vapour may be suddenly shot upwards as a vertical column of enormous height, penetrating the passing clouds. For a short distance above the vent the superheated steam sometimes exists as a transparent vapour, but it soon suffers partial condensation, forming clouds, which, if not dispersed by winds, accumulate over the mountain.

When the vapour is free from ash it forms rolling balls of fleecy cloud, but usually it carries in mechanical association more or less finely divided lava as volcanic dust and ashes, whereby it becomes yellow, brown, or even black, sometimes as foul as the densest smoke. In a calm atmosphere the dust-laden vapour may rise in immense rings with a rotatory movement, like that of vortex-rings.

Frequently the vapours, emitted in a rapid succession of jets, form cumulus clouds, or are massed together in cauliflower-like forms. The well-known " pine-tree appendage " of Vesuvius, noted by the younger Pliny in his first letter to Tacitus on the eruption in the year 79, is a vertical shaft of vapour terminating upwards in a canopy of cloud, and compared popularly with the trunk and spreading branches of the stonepine. Whilst in some cases the cloud resembles a gigantic expanded umbrella, in others it is more mushroom-shaped.

In a great eruption, the height of the mountain itself may appear dwarfed by comparison with that of the column of vapour. During the eruption of Vesuvius in April 1906, the steam and dust rose to a height of between 6 and 8 m.

At Krakatoa in 1883 the column of vapour and ashes reached an altitude of nearly 20 m.; whilst it was estimated by some authorities that during the most violent explosions the finely divided matter must have been carried to an elevation of more than 30 m.

The emission of vast volumes of vapour at high tension naturally produces much atmospheric disturbance, often felt at great distances from the centre of eruption.

ELECTRICAL EXCITEMENT

It is probably to the uprushing current of vapour that much of the electrical excitement which invariably accompanies an eruption may be referred. The friction of the steam rushing in jets through the volcanic vent must produce electrical disturbance, and indeed an active volcano has been aptly compared to a hydroelectric machine of gigantic power.

Another cause of excitement may be found in the mutual friction of the ejected cinders and ashes as they rise and fall in showers through the air. Much trituration of volcanic material may go on in the crater and elsewhere during the eruption, whereby the solid lava is reduced to a fine dust.

Other means of generating electricity are found in the chemical reactions effected in the volcano and in the sudden condensation of the emitted vapour. L. Palmieri, in the course of his investigations at the observatory on Vesuvius, found that the vapours free from cinders carried a positive charge, whilst the cinders were negative.

The electrical phenomena attending an eruption are often of great intensity and splendour. The dark ash-laden clouds of vapour are shot through and through by volcanic lightning, sometimes in rapid horizontal flashes, then in oblique forked streaks, or again in tortuous lines compared to fiery serpents, whilst the borders of the cloud may be brilliant with electric scintillations, often forming balls and stars of fire.

During the great eruption of Krakatoa remarkable phenomena were observed by ships in the Strait of Sunda, luminous balls like " St Elmo's fire " appearing at the mast-heads and the yard-arms, whilst the volcanic mud which fell upon rigging and deck was strongly phosphorescent.

Quite distinct from any electrical phenomena is that intermittent reddish glare which is often seen at night in clouds hanging over an active crater, and which is simply a glow due to reflection from ,the incandescent lava and stones in the volcanic cauldron below.

VOLCANIC RAIN AND MUD

The condensation of the vast volumes of steam exhaled during an eruption produces torrents of rain, which, mingling to a greater or less extent with the volcanic ashes, forms a hot muddy stream known in Italy as *lava d'acqua* and *lava di fango,* and in South America as *moya.* Deluges of such mud-lava may rush violently down the mountain-side and spread over the neighbouring country with terribly destructive effect, whence they are greatly dreaded by those who dwell at the base of a volcano.

The solidified volcanic mud, often mingled with larger fragments of lava, is known as *duf f* or *tufa.* Herculaneum was buried beneath a flood of mud swept down from Vesuvius during the Plinian eruption of 79, and the hard tufaceous crust which thus sealed up the illfated city came in turn to be covered by lava-flows from subsequent eruptions: hence the difficulty of excavating at Herculaneum compared with similar work at Pompeii, where there was probably much less mud, since the city, having been at a greater distance from the volcanic centre, was overwhelmed in great measure by loose ashes, capable of removal with comparative ease.

It sometimes happens that volcanic mud is formed by the mingling of hot ashes not directly with rain but with water from streams and lakes, or even, as in Iceland, with melted snow. A torrent of mud was one of the earliest symptoms of the violent eruption of Mont Pele in Martinique in 1902. This mud had its source in the Etang Sec, a crater-basin high up on the S.W. side of the mountain.

By the explosive discharge of ashes and vapours mingled with the water of the tarn there was produced a vast volume of hot muddy matter which on the 5th of May suddenly escaped from the basin, when a huge torrent of boiling black mud, charged with blocks of rock and moving with enormous rapidity, rolled like an avalanche down the gorge of the Riviere Blanche. If a stream of lava obstructs the drainage of a volcano, it may give rise to floods.

EJECTED BLOCKS

When a volcano after a long period of repose starts into fresh activity, the materials which have accumulated in the crater, including probably large blocks from the disintegration of the crater-walls, have to be ejected. If the lava from the last eruption has consolidated as a plug in the throat of the volcano, the conduit may be practically closed, and hence the first effort of the renewed activity is to expel this obstruction.

The hard mass becomes shattered by the explosions, and the angular fragments so formed are hurled forth by the outrushing stream of vapour. When the discharge is violent, the vapour, as it rushes impetuously up the volcanic duct, may tear fragments of rock from its walls and project them to a considerable distance from the vent. Such ejected blocks, by no means uncommon in the early stages of an eruption, are often of large size and naturally vary according to the character of the rocks through which the duct has been opened.

They may be irregular masses of igneous rocks, possibly lavas of earlier eruptions, or they may be stratified, sedimentary and fossiliferous rocks representing the platform on which the volcano has been built, or the yet more deeply seated fundamental rocks. By Dr H.

J. Johnston-Lavis, who specially studied the ejected blocks of Vesuvius, the volcanic materials broken from the cone are termed " accessory " ejecta, whilst other fragmentary materials he conveniently calls " accidental " products, leaving the term " essential " ejecta for plastic lava, ashes, crystals, &c.

Masses of Cretaceous or Apennine limestone ejected from Somma are scattered through the tuffs on the slopes of Vesuvius; and objects carved in such altered limestone are sold to tourists as " lava " ornaments. Under the influence of volcanic heat and vapours, the ejected blocks suffer more or less alteration, and may contain in their cavities many crystallized minerals. Certain blocks of sandstone ejected occasionally at Etna are composed of white granular quartz, permeated with vitreous matter and encased in a black scoriaceous crust of basic lava.

A rock consisting of an irregular aggregation of coarse ejected materials, including many large blocks, is known as a " volcanic agglomerate." Any fragmental matter discharged from a volcano may form rocks which are described as " pyroclastic." *Cinders, Ashes and* Dust. - After the throat of a volcano has been cleared out and a free exit established, the copious discharge of vapour is generally accompanied by the ejection of fresh lava in a fragmentary condition.

If the ejected masses bear obvious resemblance to the products of the hearth and the furnace, they are known as " cinders " or " scoriae," whilst the small cinders not larger than walnuts often pass under their Italian name of "lapilli". When of globular or ellipsoidal form, the ejected masses are known as " bombs " or " volcanic tears." Other names are given to the smaller fragments.

If the lava has become granulated it is termed " volcanic sand "; when in a finer state of division it is called ash, or if yet more highly comminuted it is classed as dust; but the latter terms are sometimes used interchangeably.

The pulverized material, consisting of lava which has been broken up by the explosion, or triturated in the crater, is often discharged in prodigious quantity, so that after an eruption the country for miles around the volcano may be covered with a coating of fine ash or dust, sometimes nearly white, like a fall of snow, but often of greyish colour, looking rather like Portland cement, and in many cases becomirg reddish by oxidation of the ferruginous constituents. Even when first ejected the ash is sometimes cocoa-coloured.

This finely divided lava insinuates itself into every crack and cranny, reaching the interior of houses even when windows and doors are closed. A heavy fall of ash or cinders may cause great structural damage, crushing the roofs of buildings by sheer weight, as was markedly the case at Ottajano and San Guiseppe during the eruption of Vesuvius in April 1906.

On this occasion the dry ashes slipped down the sides of the volcanic cone like an avalanche, forming great ashslides with ridges and furrows rather like barrancos, or ravines, caused by rain. The burial of Ottajano and San Ginseppe in 1906 by Vesuvian ejecta, mostly lapilli, has been compared with that of Pompeii in 79.

Deposits of volcanic sand and ashes retain their heat long after ejection, so that rain will cause them to evolve steam, and if the rain be heavy and sudden it may produce explosions with emission of great clouds of vapour.

The fall of ash is at first prejudicial to vegetation, and is often accompanied or followed by acid rain; but ultimately the ash may prove beneficial to the soil, chiefly in consequence of the alkalis which it contains. The " May dust " of Barbados was a rain of volcanic ash which fell in May 1812 from the eruption of the Soufriere in St Vincent. It is estimated that the amount of dust which during this eruption fell on the surface of Barbados, Ioo m. distant from the eruptive centre, was about 3,000,000 tons.

The distance to which ash is carried depends greatly on the atmospheric conditions at the time of the eruption. Ashes from Vesuvius in an eruption in the year 472 were carried, it is said, as far as Constantinople.

During an eruption of Cotopaxi, on the 3rd of July 1880, observed by E. Whymper, an enormous black column of dust-laden vapour was shot vertically upwards with such rapidity that in less than a minute it rose to a height estimated at 20,000 ft. above the crater-rim, or nearly 40,000 ft. above sea-level, when it was dispersed by the wind over a very wide area.

It is believed that the amount of dust in this discharge must have been more than 2,000,000 tons. Enormous quantities of dust ejected from Krakatoa in 1883 were carried to prodigious distances, samples having been collected at more than a thousand miles from the volcano; whilst the very fine material in ultramicroscopic grains which remained suspended for months in the higher regions of the atmosphere seems to have enjoyed an almost world-wide distribution, and to have been responsible for the remarkable sunsets at that period.

The ash falling in the immediate vicinity of a volcanic vent will generally be coarser than that carried to a distance, since the particles as they are wafted through the air undergo a kind of sifting. Professor J. W. Judd, who made an exhaustive examination of the products of the eruption of Krakatoa, found that the dust near the volcano was comparatively coarse, dense and rather darkcoloured, in consequence of the presence of numerous fragments of heavy, dark, crystalline minerals, whilst the dust at a distance was excessively fine and perfectly white. According to this observer, the particles tended to fall in the following order: magnetite, pyroxenes, felspar, glass.

The finely comminuted material, carried to a great height in the atmosphere, consisted largely of delicate threads and attenuated plates of vitreous matter, in many cases hollow and containing air-bubbles. The greater part of the dust was formed by the mutual attrition of fragments of brittle pumice as they rose and fell in the crater, which thus became a powerful "dust-making mill."

By this trituration of the pumiceous lava, carried on for a space of three months during which the eruption lasted, the quantity of finely pulverized material must have been enormous; yet the amount of ejected matter was probably very much less than that extruded during some other historical eruptions, such as that of Tomboro in Sumbawa, in 1815.

The explosions at Krakatoa were, however, exceptionally violent, having been sufficient to project some of the finely pulverized lava to an altitude estimated to have been at least 30 m. It is usually impossible during a great eruption to determine the height of the column of "smoke" since it hangs over the country as a pall of darkness.

The great black cloud, which was so characteristic a feature in the terrible eruptions in the West Indies in 1902, was formed of steam with sulphur dioxide and other gases, very heavily charged with incandescent sand or dust, forming a dense mixture that in some respects behaved like a liquid. Unlike the Krakatoa dust, which was derived from a vitreous pumice, the solid matter of the black cloud was largely composed of fragments of crystalline minerals.

According to Drs Anderson and Flett it is not impossible that on the afternoon of the 17th of May 1902, the solid matter ejected from the Soufriere of St Vincent amounted to several billions of tons, and that some of the dust fell at distances more than 2000 m. east of the centre of eruption.

LAVA

The volcanic cinders, sand, ashes and dust described above are but varied forms of solidified lava. Lava is indeed the most characteristic product of volcanic activity. It consists of mineral matter which is, or has been, in a molten state; but the liquidity is not due to simple dry fusion.

The magma, or subterranean molten matter, may be regarded as composed essentially of various silicates, or their constituents, in a state of mutual solution, and heavily charged with certain vapours or gases, principally water-vapour, superheated and under pressure. In consequence of the peculiar constitution of the magma, the order in which minerals separate and solidify from it on cooling does not necessarily correspond with the inverse order of their relative fusibility.

The lava differs from the magma before eruption, inasmuch as water and various volatile substances may be expelled on extrusion. The rapid escape of vapour from the lava contributes to the explosive phenomena of an eruption, whilst the rate at which the vapour is disengaged depends largely on the viscosity of the magma.

The lava on its immediate issue from the volcanic vent is probably at a white heat, but the temperature is difficult of determination since the molten matter is usually not easy of approach, by reason of the enshrouding vapour.

Determinations of temperature are generally made at a short distance from the exit, when the lava has undergone more or less cooling, or on a small stream from a subordinate vent. A. Bartoli, using a platinum electric resistance pyrometer, found that a stream of lava near a *bocca,* or orifice of emission, on Etna, in the eruption of 1892, had at a depth of one foot a temperature of 1060° C. In the lavas of Vesuvius and Etna thin wires of silver and of copper have frequently been melted.

Probably the lava at the surface of the stream has a temperature of something like 110o° C., but this must not be assumed to be its temperature at the volcanic focus. C. Doelter, in some experiments on the melting-point of lava by means of an electric furnace, found that a lava from Etna softened at from 962° to 970° C. and became fluid at 1010° to 1040°, whilst a Vesuvian lava softened at 1030° to 1060° and acquired fluidity at 1080° to 1090°.

These results were obtained at ordinary atmospheric pressure, but it has been assumed that the melting-point of lava at a great depth would, through pressure alone, exceed that obtained in the laboratory. On the other hand the presence of water and of certain volatile fluxes in the magma lowers the fusing-point, and hence the extruded lava from which these have largely escaped may be much less fusible than the original magma.

It should be noted that all determinations of the melting-points of minerals and rocks involving ocular inspection of the physical state of the material are liable to considerable error, and the only accurate method seems to be that of determining the point at which absorption of heat abruptly occurs - the latent heat of fusion. It is believed that the temperature of lava in the volcanic conduit may be in some cases sufficiently high to fuse the neighbouring rocks, and so melt out a passage through them in its ascent. The wallrock thus dissolved in the magma will not be without influence on the composition of the lava with which it becomes assimilated.

Many interesting observations are on record with regard to the heating effect of lava on metals and other objects with which it may have come in contact. Thus, after the destruction of Torre del Greco by a current of lava from Vesuvius in 1794, it was found that brass in the houses under the lava had suffered decomposition, tje copper having become crystallized; whilst silver had been not only fused but sublimed. This indicates a temperature of upwards of 1000° C. Panes of glass in the windows at Torre del Greco on the same occasion suffered devitrification.

Notwithstanding the high temperature of lava on emission, it cools so rapidly, and the consolidated lava conducts heat so slowly, that vegetable structures may be involved in a lava-flow without being entirely destroyed. A stream of lava on entering a wood, as in the sylvan region on Etna, may burn up the undergrowth but leave many of the larger trees with their trunks merely carbonized. On Vesuvius a lava-flow has been observed to surround trees while the foliage has been apparently uninjured.

A vertical trunk of a coniferous tree partially enveloped in Tertiary basalt occurs at Gribon in the Isle of Mull, as described by Sir A. Geikie and others; plant-remains in basalt from the Bo'ness coalfield in Linlithgowshire have been noticed by H. M. Cadell; and attention has been called by B. Hobson to a specimen of scoriaceous basalt, from Mexico, which shows the impression of ears of maize and even relics of the actual grains.

In consequence of the slow transmission of heat by solid lava, the crust on the surface of a stream may be crossed with impunity whilst the matter is still glowing at a short distance below. Lichens may indeed grow on lava which remains highly heated in the interior.

The solidified surface of a sheet of lava may be smooth and shining, sometimes quite satiny in sheen, though locally wrinkled and perhaps even ropy or hummocky, the irregularities being mainly due to superficial movement after partial solidification. The " corded lava " has a surface similar to that often seen on blastfurnace slag, and is suggestive of a tranquil flow.

After a lava stream has become crusted over on cooling, the subjacent lava, still moving in a viscous condition, tends to tear the crust, forming irregular blocks, or clinkers, which are carried forward by the flow and ultimately left in. the form of confused heaps, perhaps of considerable magnitude.

The front of a stream may present a wall of scoriaceous fragments looking like a huge pile of coke. As the clinkers are carried along, on the surface of the lava, they produce by mutual friction a crunching noise; and the sluggish flow of the lava-stream laden with its burden has been compared with that of a glacier.

Since the upper part of the stream moves more rapidly than the lower, which is retarded by cooling in contact with the bed-rock, the superficial clinkers are carried forward and, rolling over the end, may become embedded in the lava as it advances. Scoriae formed on the top of a stream may thus find their way to the base.

Rockfragments or other detrital matter occurring in the path of the lava will be caught up by the flow and become involved in the lower part of the molten mass; whilst the rocks over which the lava travels may suffer more or less alteration by the heat of the stream.

The rapidity of a lava flow is determined partly by the slope of the bed over which it moves and partly by the consistency of the lava, this being dependent on its chemical composition and on the conciitions of cooling. In an eruption of Mauna Loa, in Hawaii, in 1855, the lava was estimated to flow at a rate of 40 m. an hour; and at an eruption of Vesuvius in 1805 a velocity of more than 50 m. an hour, at the moment of emission, was recorded.

The rapidity of flow is, however, rapidly checked as the stream advances, the retardation being very marked in small flows. Where lava travels down a steep incline there is naturally a great tendency to form a rugged surface, whilst a quiet flow over a flat plane favours smoothness. If the lava meet a precipice it may form a cascade of great beauty, the clinkers rapidly rolling down with a clatter, as described by Sir W. Hamilton in the eruption of Vesuvius in 1771, when the fiery torrent had a perpendicular fall of 50 ft.

In Hawaii the smooth shining lava, often superficially waved and lobed, is known as *pahoehoe,* whilst the rugged clinker beds are termed aa. These terms are now used in general terminology, having been introduced by American geologists.

The fields of aa often contain lava-balls and bombs. It may be said that the pahoehoe corresponds practically with the *Fladen lava* of German vulcanologists, and the as with their *Schollen lava.* Rugged flows are known in Auvergne as *cheires.* The surface of a clinker-field has often a horribly jagged character, being covered with ragged blocks bristling with sharp points. In the case of an obsidian-flow a most dangerous surface is produced by the keen edges and points of the fragmentary volcanic glass.

If, after a stream of lava has become crusted over, the underlying magma should flow away, a long cavern or tunnel may be formed. Should the flow be rapid the roof may collapse and the fragments, falling on to the stream, may be carried forward or become absorbed in the fused mass. The walls and roof of a lava-cave are occasionally adorned with stalactites, whilst the floor may be covered with stalagmitic deposits of lava.

The volcanic stalactites are slender, tubular bodies, extremely fragile, often knotted and rippled. Beautiful examples of lava stalactites from Hawaii have been described by Professor E. S. Dana. Caverns may also be formed in lava-flows by the presence of large bubbles, or by the union of several bubbles. It may happen, too, that certain monticules thrown up on the surface of the lava are hollow, of which a famous example is furnished by the Caverne de Rosemond, at the base of Piton Barry, in the Isle of Reunion.

It is of great interest to determine whether molten lava contracts or expands on solidification, but the experimental evidence on this subject is rather conflicting. According to some observers a piece of solid lava thrown on to the surface of the same lava in a liquid state will sink, while according to others it floats.

It has often been observed that cakes formed by the natural fracture of the crust on the lava of Kilauea sink in the liquid mass, but it has been suggested that the fragments are drawn down by convection-currents.

On the other hand a solid piece, though denser than the corresponding liquid, may be buoyed up for a time by the viscous condition of the molten lava. Moreover, the presence of minute vesicles may lighten the mass. Although the minerals of a rock-magma may separately contract on crystallization it does not follow that the magma itself, in which they probably exist in a state of solution, will undergo on crystallization a similar change of volume. On the whole, however, there seems reason to believe that lava on solidifying almost always diminishes in volume and consequently increases in density.

According to the experiments of C. Doelter the specific gravity of molten lava is invariably less than that of the same lava when solid, though in some cases the difference is but slight. In a vitreous or isotropic condition the lava has a lower density than when crystalline.

Experiments by Dr C. Barus showed that a diabase of specific gravity 3.017 formed a glass of sp. gr. 2.717, and melted to a liquid of sp. gr. 2.52. J. A. Douglas on examining various igneous rocks found that in all cases the rock in a vitreous state had a lower sp. gr. than in a crystalline condition, the difference being greatest in the acid plutonic rocks. A. Harker, however, has called attention to the fact that the glassy selvage of certain basic dykes in Scotland is denser than the same rock in a crystalline condition in the interior of the dykes.

PHYSICAL STRUCTURE OF LAVAS

An amorphous vitreous mass may result from the rapid cooling of a lava on its extrusion from the volcanic vent. The common type of volcanic glass is known as obsidian. Microscopic examination usually shows that even in this glass some of the molecules of the magma have assumed definite orientation, forming the incipient crystalline bodies known as microlites, &c. By the increase of these minute enclosures, in number and magnitude, the lava may become devitrified and assume a lithoidal or stony structure.

If the molten magma consolidate slowly, the various silicates in solution tend to separate by crystallization as their respective points of saturation are reached. Should the process be arrested before the entire mass has crystallized, the crystals that have been developed will be embedded in the residual magma, which may, on consolidation, form a vitreous base.

It is believed that in many cases the lava brings up, through its conduit, myriads of crystals that have been developed during slow solidification in the heart of the volcanic apparatus. Showers of crystals of leucite have occurred at Vesuvius, of labradorite at Etna, and of pyroxene at Vesuvius, Etna and Stromboli.

These " intratelluric crystals were probably floating in the molten magma, and had they remained in suspension, this magma might on consolidation have enveloped them as a ground-mass or base. A rock so formed is generally known as a " porphyry," and the structure as porphyritic. In such a lava the large crystals, or phenocrysts, evidently represent an early phase of consolidation, and the minerals of the matrix a later stage.

It is notable that the intratelluric crystals often lack sharpness of outline, as though they had suffered corrosion by attack of the molten magma, whilst they may contain vitreous enclosures, suggesting that the surrounding mass was liquid during their consolidation.

It is believed that the more slowly consolidation has occurred, the larger generally are the crystals; and the higher the temperature of the magma the greater the corrosion or resorption. Possibly under certain conditions the phenocrysts and the ground-mass may have solidified simultaneously.

In some cases the entire igneous mass assumes a crystalline structure, or becomes " holocrystalline." Such a structure is well displayed when the magma has consolidated at considerable depths, cooling slowly under great pressure, and forming rocks which are termed " plutonic " or " abyssal " to distinguish them from rocks truly volcanic, or those which, if not effusive, like lava-flows, have at least solidified very near to the surface as dykes and sills.

Volcanic and plutonic rocks pass, however, into each other by gradual transition. The dyke-rocks, or intrusive masses, form an intermediate group sometimes distinguished under the name of " hypabyssal " rocks, as suggested by W. C. Brogger. Lavas extruded in submarine eruptions may have solidified under a great weight of seawater, and therefore to that extent rather under plutonic conditions.

CHEMICAL COMPOSITION OF LAVAS

Lavas are usually classified roughly, from a chemical point of view, in broad groups according to the proportion of silica which they contain. Those in which the proportion of silica reaches 66% or upwards are said to be acid or acidic, whilst those in which it falls to 55% or below are called basic lavas.

The two series are connected by a group of intermediate composition, whilst a small number of igneous rocks of exceptional type are recognized as ultrabasic. Professor F. W. Clarke has suggested a grouping of igneous rocks as per-silicic, medio-silicic and subsilicic, in which the proportion of silica is respectively more than 60, between 50 and 60, or less than 50%.

By far the greater part of all lavas consists of various silicates, either crystallized as definite minerals or unindividualized as volcanic glass. In addition, however, to the mineral silicates, a volcanic rock may contain a limited amount of free acid and basic oxides, represented by such minerals as quartz and magnetite.

Rhyolite may be cited as a typical example of an acid lava, andesite as an intermediate and basalt as a basic lava. The various volcanic rocks are described under their respective headings, so that it is needless to refer here to their chemical or mineralogical composition.

In the course of the life of a volcano, the lava which it emits may undergo changes, within moderate limits, being at one time more acid, at another more basic. Such changes are sometimes connected with a shifting of the axis of eruption. Thus at Etna the lavas from the old axis of Trifoglietto in the Valle del Bove were andesites, with about 55% of silica, but those rising in the present conduit are doleritic, with a silica-content of only about 50%.

It seems probable that, to a limited extent, changes in the character of a lava may sometimes be due to contact of the magma with different rocks underground: if these are rich in silica, the acidity of the lava will naturally increase; while if they are rich in calcareous and ferromagnesian constituents, the basicity will increase: the variation is consequently apt to be only local, and probably always slight.

By von Richthofen and some others it has been held that during a long period of igneous activity a definite order in the succession of the erupted rocks is everywhere constant; but though some striking coincidences may be cited, it can hardly be said that this generalization has been satisfactorily established.

It has, however, often been observed, as emphasized by Professor Iddings, that a volcanic centre will start with the emission of lavas of neutral or intermediate type, followed in the course of a geological period by acid and basic lavas, and ending with those of extreme composition, indicating progressive change in the magma.

The old idea of a universal magma, or continuous pyrosphere, has been generally abandoned. Whatever may have been the case in a primitive condition of the interior of the earth, it seems necessary to admit that the magma must now exist in separate reservoirs.

The independent activity of neighbouring volcanoes strikingly illustrated in Kilauea and Mauna Loa in Hawaii, only 20 m. apart, suggests a want of communication between the conduits; and though the lavas are very similar at these two centres, it would seem that they can hardly be drawn from a common source.

Again, the volcanoes of southern Italy and the neighbouring islands exhibit little or no sympathy in their action, and emit lavas of diverse type. The lavas of Vulcano, one of the Lipari Isles, are rhyolitic, whilst those of Stromboli, another of the group, are basaltic.

It is believed that the magma in a subterranean reservoir, though originally homogeneous, may slowly undergo certain changes, whereby the more basic constituents migrate to one quarter whilst the acid segregate in another, so that the canal, at successive periods, may bring up material of different types.

It has often been observed that all the rocks from a definite igneous centre have a general similarity in chemical and mineralogical characters. This relationship is called, after Professor Iddings, " consanguinity," and appears to be due to the fact that the rocks are drawn from a common source.

Professor Judd pointed out the existence of distinct " petrographical provinces," within which the eruptive rocks during a given geological period have a certain family likeness and have appeared in definite succession. Thus he recognized a Brito-Icelandic petrographical province of Tertiary and recent lavas. It has been shown by A. Harker that alkali igneous rocks are generally associated with the Atlantic type of coast-line and sub-alkali rocks with the Pacific type.

Although changes in the character of an erupted product from a given centre are usually brought about very slowly, it has often been supposed that even in the course of a single prolonged eruption, or series of eruptions, the character of the lava may vary to some extent. That this is not, however, usually the case has been repeatedly proved. M. H. Arsandaux, for instance, analysed the bombs of augite-andesite thrown out from Santorin at the beginning of the eruption of 1866, others ejected in 1867, and others again at the close of the eruption in 1868; and he found no important variation in the composition of the magma during these successive stages.

Moreover, Professor A. Lacroix found that the material extruded from Vesuvius in 1906 remained practically of the same composition from the beginning to the end of the eruption, and further, that it presented great analogy to that of 1872 and even to that of 1631.

All the Vesuvian lavas are of the type of rock known as leucotephrite or leucitetephrite, or they pass, by the presence of a little olivine, into leucite-basanite. Leucite is characteristic of the lavas of Vesuvius, whilst it is excessively rare in those of Etna, where a normal doleritic type prevails. Nepheline, a felspathoid related to leucite, is characteristic of certain lavas, such as those of the Canary Islands, which comprise nepheline-tephrites and nepheline-basanites.

Most of the lavas from the volcanoes of South America consist of hypersthene-andesite, and it is notable that the fragmental ejectamenta from the eruptions of St Vincent and Martinique in 1902 and from Krakatoa in 1883 were evidently derived from a magma of this Pacific type.

It commonly happens that acid lavas are paler in colour, less dense and less fusible than basic lavas, and they are probably drawn in some cases from shallower depths. As a consequence of the ready fusibility of many basic lavas, they flow freely on emission, running to great distances and forming far-spreading sheets, whilst the more acid lavas rapidly become viscid and tend to consolidate nearer to their origin, often in hummocky masses.

The shape of a volcanic mountain is consequently determined to a large extent by the chemical character of the lavas which it emits. In the Hawaiian Islands, for instance, where the lavas are highly basic and fluent, they form mountains which, though lofty, are flat domes with very gently sloping sides.

Such is the fluidity of the lava on emission that it flows freely on a slope of less than one degree. In consequence, too, of this mobility, it is readily thrown into spray and even projected by the expansive force of vapour into jets, which may rise to the height of hundreds of feet and fall back still incandescent, producing the appearance of "fire fountains." The emission is not usually accompanied, however, by violent explosions, such as are often associated with the eruption of magmas of less basic and more viscous nature.

The viscosity of the lava at Kilauea was estimated by G. F. Becker to be about fifty times as great as that of water. It may be pointed out that the fusibility of a lava depends not on the mere fact that it is basic, but rather on the character of the bases. A lava from Etna or Vesuvius may be really as basic as one from Hawaii.

Capillary Lava

A filamentous form of lava well known at Kilauea, in Hawaii, is termed *Pele's hair*, after Pele, the reputed goddess of the Hawaiian volcanoes. It resembles the capillary slag much used in the arts under the name of " mineral wool " - a material formed by injecting steam into molten slag from an iron blast-furnace. It is commonly supposed that Pele's hair has been formed from drops of lava splashed into the air and drawn out by the wind into fine threads.

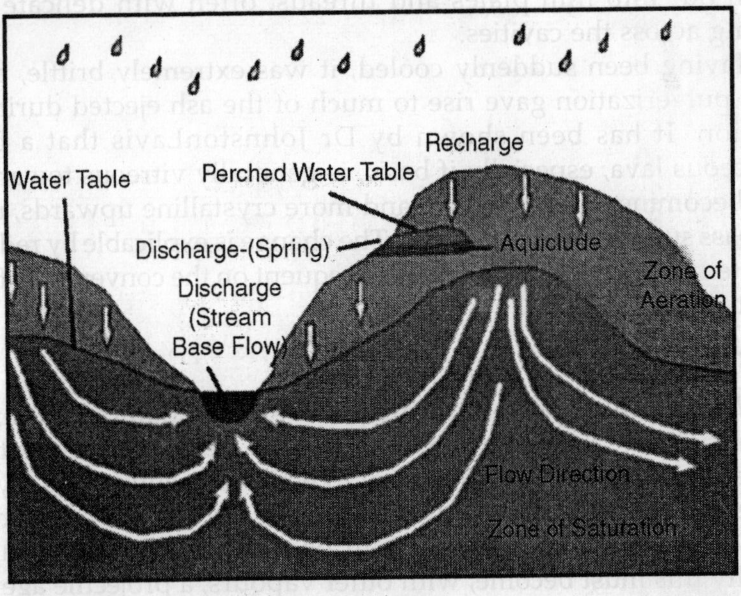

Fig. Capillary Lava

According, however, to Major C. E. Dutton, the filaments are formed on the eddying surface of the lava by the elongation of minute vesicles of water-vapour expelled from the magma. C. F. W. Krukenberg, who examined the hair microscopically, figured a large number of fibres, some of which showed the presence of minute vesicles and microscopic crystals, the former when drawn out rendering the thread tubular.

In a spongy vitreous scoria from Hawaii, described as " thread-lace," a polygonal network of delicate fibres forms little skeleton cells. Capillary lava is not confined to the Hawaiian volcanoes: it is known, for example, in Reunion, and may be formed even at Vesuvius.

Pumiceous Lava

The copious disengagement of vapour in a glassy lava gives rise to the light cellular or spongy substance, full of microscopic pores, known as pumice. It is usually, though not invariably, produced from an acid lava, and may sometimes be regarded as the solidified foam of an obsidian.

During the eruption of Krakatoa in 1883 enormous quantities of pumice were ejected, and were carried by the sea to vast distances, until they ultimately became water-logged and sank. Professor Judd found the pumice to consist of a vitreous lava greatly inflated by imprisoned vapours; the walls of the air-cells were formed of the lava drawn out into thin plates and threads, often with delicate fibres running across the cavities.

Having been suddenly cooled, it was extremely brittle, and its ready pulverization gave rise to much of the ash ejected during this eruption. It has been shown by Dr JohnstonLavis that a bed of pumiceous lava, especially if basic, is generally vitreous towards the base, becoming denser, darker and more crystalline upwards, until it may pass superficially into scoria. The change is explicable by reduction in the temperature of the magma consequent on the conversion of water into steam.

Water in Lavas

Whether an eruption is of an explosive or a tranquil character must depend largely, though not wholly, on the chemical composition of the magma, especially on the extent to which it is aquiferous. By relief of pressure on the rise of the column in the volcanic channel, or otherwise, more or less steam will be disengaged, and if in large quantity this must become, with other vapours, a projectile agency of enormous power. The precise physical condition in which water exists in the magma is a matter of speculation, and hence Johnston-Lavis proposed to designate it simply as H_2O. Water above its critical point, which is about 370° C. or 698°F., cannot exist as a liquid, whatever be the pressure, neither is it an ordinary vapour. It has been estimated that the critical point would probably be reached at a depth of about 7 m. At very high temperatures the elements of water may exist in a state of dissociation.

Much discussion has arisen as to the origin of the volcanic water, but probably it is not all attributable to a single source. Some may be of superficial origin, derived from rain, river or sea; whilst the upward passage of lava through moist strata must generate large volumes of steam. It has often been remarked that wet weather increases the activity of a volcano, and that in certain mountains the eruptions are more frequent in winter.

According, however, to Professor A. Ricco's prolonged study of Etna, rain has no apparent influence on the activity of this mountain, and indeed the number of eruptions in winter, when rains are abundant, seems rather less than in summer.

The popular belief that explosive action is due to the admission of water to the volcanic focus is founded mainly on the topographic relation of volcanoes to large natural bodies of water, many being situated near the shore of a continent or on islands or even on the sea-floor.

Salt water gaining access to heated rocks, through fissures or by capillary absorption, would give rise not only to watervapour but to the volatile chlorides so common in volcanic exhalations.

Yet it is notable that comparatively little chlorine is found among the products exhaled by the volcanoes of Hawaii, though these are typically insular. L. Palmieri, however, described certain sublimates on lava at Vesuvius after the eruption of 1872 as deposits of "sea-salt," to show that they were not simply sodium chloride, but contained other constituents found in sea-water.

It has been supposed that water on reaching the hot walls of a subterranean cavity would pass into the spheroidal state, and on subsequent reduction of temperature might come into direct contact with the heated surface, when it would flash with explosive violence into steam.

Such catastrophes probably occur in certain cases. When, for example, a volcano becomes dormant, water commonly accumulates in the crater, and on a renewal of activity this craterlake may be absorbed through fissures in the floor leading to the reopened duct, and thus become rapidly, even suddenly, converted into vapour. But such incidents are accidental rather than normal, and seem incompetent to account for volcanic activity in general. When a stream of lava flows into the sea it no doubt immediately generates a prodigious volume of steam; but this is only a temporary phenomenon, for the lava rapidly becomes chilled by the cold water, with formation of a superficial solid layer, which by its low thermal conductivity allows the internal mass to cool slowly and quietly. In the great eruption of Krakatoa in 1883 the sea-water gained occasional access to the molten lava, and by its cooling effect checked the escape of vapour, thus temporarily diminishing the volcanic activity. But Judd compares this action to that of fastening down the safety-valve of a steam-boiler.

The tension of the elastic fluids being increased by this repression would give rise subsequently to an explosion of greater violence; and hence the short violent paroxysms characteristic of the Krakatoa eruption were due to what he calls a " check and rally " of the subterranean forces. The action in the volcanic conduit has, indeed, been compared with that of a geyser.

The downward passage of water through fissures must be confined to the upper portion of the earth's crust known as the " zone of fracture," for it is there only that open channels can exist. Water might also percolate through the pores of the rocks, but even the pores are closed at great depths. It was shown many years ago by G. A. Daubree that water could pass to a limited extent through a heated rock against the pressure of steam in the opposite direction. According to S. Arrhenius, water may pass inwards through the sea-bottom by osmotic pressure.

As the melting points of various silicates are lowered by admixture with water, it appears that the access of surface-waters to heated rocks must promote their fusibility. Judd has suggested that the proximity of large bodies of water may be favourable to volcanic manifestations, because the hydrated rocks become readily melted by internal heat and thus yield a supply of lava.

Whilst some of the water-vapour exhaled from a volcano is undoubtedly derived from superficial sources, notably in such insular volcanoes as Stromboli, the opinion has of late years been gaining ground, through the teaching of Professor E. Suess and others, that the volcanic water must be largely referred to a deep-seated subterranean origin - that it is, in a word, " hypogene " or magmatic rather than meteoric.

It is held that the magma as it rises through the volcanic conduit brings up much water-vapour and other gaseous matters derived from original sources, perhaps a relic of what was present in the earth in its molten condition, having possibly been absorbed from a dense primordial atmosphere, or, as suggested by Professor T. C. Chamberlin, entrapped by the globe during its formation by accretion of planetesimal matter.

Water brought from magmatic depths to the surface, and appearing there for the first time, has been termed " juvenile," and it has been assumed that such water may be seen in hot springs like those at Carlsbad. Professor J. W. Gregory has suggested that certain springs in the interior of Australia may derive part of their supply from juvenile or plutonic waters.

According to A. Gautier, the origin of volcanic water may be found in the oxidation of hydrogen, developed from masses of crystalline rock, which by subsidence have been subjected to the action of subterranean heat.

Volcanic Vapours

It seems not unlikely that the vapours and gases exist in the volcanic magma in much the same way that they can exist in molten metal. It is a familiar fact that certain metals when melted can absorb large volumes of gases without entering into chemical combination with them. Molten silver, for example, is capable of absorbing from the atmosphere more than twenty times its volume. of oxygen, which it expels on solidification, thus producing what is called the " spitting of silver." Platinum again can absorb and retain when solid, or occlude, a large volume of hydrogen, that can be expelled by heating the metal in vacuo.

In like manner molten rock under pressure can absorb much steam. It appears that many igneous rocks contain gases locked up in their pores, not set free by pulverization, yet capable of expulsion by strong heat.

Tilden has found that granite, gabbro, basalt and certain other igneous rocks enclose many times their volume of gases, chiefly hydrogen and carbon dioxide, with carbon monoxide, methane and nitrogen.

Thus, the basalt of Antrim in Ireland, which is a Tertiary lava, yielded eight times its volume of gas having the following percentage composition: hydrogen 36.15, carbon dioxide 32.08, carbon monoxide 20.08, methane 10, nitrogen. No doubt some of the gases evolved on heating rocks may be generated by reactions during the experiment, as shown by M. W. Travers, and also by Armand Gautier.

It has been pointed out by Gautier that the gas exhaled from Mont Pele during the eruption of 1902 had practically the same composition as that which he obtained on heating granite and certain other rocks.

According to this authority a cubic kilometre of granite heated to redness would yield not less than 26,000,000 tons of water-vapour, besides other gases. If then a mass of granite in the earth's crust were subject to a great local accession of heat it might evolve vast volumes of gaseous, matter, capable of producing an eruption of explosive type. Judd found that the little balls of Siberian obsidian called marekanite threw off, when strongly heated, clouds of finely divided particles formed by rupture of the distended mass through the escape of vapour. Pitchstone when ignited loses in some cases as much as to % of its weight, due to expulsion of water.

Much of the steam and other vapour brought up from below by the lava may be evolved on mere exposure to the air, and hence a stream freshly extruded is generally beclouded with more or less vapour. Gaseous bubbles in the body of the lava render it vesicular,. especially in the upper part of a stream, where the pressure is relieved, and the vesicles by the onward flow of the lava tend to become elongated in the direction of movement.

Vesiculation, being naturally resisted by cohesion, is not common in very viscid lavas of acid type, nor is it to be expected where the lava has been subject to great pressure, but it is seen to perfection in surface-flows of liquid lavas of basaltic character.

A vesicular structure may sometimes be seen even in dykes, but the cavities are usually rounded rather than elongated, and are often arranged in bands parallel to the walls of the dyke. A very small proportion of water in a lava. suffices to produce vesiculation. Secondary minerals developed in a cellular lava may be deposited in the steam-holes, thus producing an amygdaloidal rock.

After the surface of a lava-stream has become crusted over, vapour may still be evolved in the interior of the mass, and in seeking release may elevate or even pierce the crust. Small cones may thus be thrown up on a lava-flow, and when vapour escapes from terminal or lateral orifices they are known as " spiracles."

The steam may issue with sufficient projectile force to toss up the lava in little fountains. When the lava is very liquid, as in the Hawaiian volcanoes, it may after projection from the blow-hole fall back in drops and plastic clots, which on consolidation form, by their union, small cones.

Vapour-vents on lava are often known as fumaroles. The character of the gaseous exhalations varies with the temperature, and the following classification was suggested by C. Sainte-Claire Deville:

- Dry or white fumaroles having a temperature above 500° C. and evolving compounds of chlorine, and perhaps fluorine.

- Acid fumaroles, exhaling much steam, with hydrocholoric acid and sulphur dioxide.
- Alkaline fumaroles, at a temperature of about 100°, with much steam and ammonium chloride and some sulphuretted hydrogen.
- Cold fumaroles, below100°, with aqueous vapour, carbon dioxide and sulphuretted hydrogen.
- Mofettes, indicating the expiring phase of vulcanism.

A similar sequence of emanations, following progressive cooling of the lava, has been noted by other observers. During an eruption, the gaseous products may vary considerably. Johnston-Lavis found at Vesuvius that the vapour which first escaped from the boiling lava contained much sulphurous acid, and that hydrochloric acid and other chlorides. appeared later.

If the vapours exhaled from volcanoes were derived originally from superficial sources, the lava would, of course, merely return to the surface of the earth what it had directly or indirectly absorbed.

But if, as is now rather generally believed, much if not most of the volcanic vapour is derived from original subterranean sources, it must form a direct contribution from the interior of the earth to the atmosphere and hydrosphere, and consequently becomes of extreme geological interest.

DESCRIPTION OF SPECIAL GASES AND VAPOURS

Hydrochloric acid, HCI, escapes abundantly from many vents, often accompanied with the vapours of certain metallic chlorides, and is responsible for much of the acrid effects of volcanic exhalations. To avoid dangerous vapours an active volcano should be ascended on the windward side.

Free hydrofluoric acid, HF, has sometimes been detected with the hydrochloric acid among Vesuvian vapours, and silicon fluoride, SiF 4, has also been reported. Sulphuretted hydrogen, H_2 S, is a frequent emanation, and being combustible may contribute to the lambent flames seen in some eruptions.

It readily suffers oxidation, giving rise to sulphur dioxide and water. By the interaction of hydrogen sulphide and carbon dioxide, water and carbon oxysulphide, COS, are formed; whilst by reaction with sulphur dioxide, water and free sulphur are produced, such being no doubt the origin of many deposits of volcanic sulphur.

Hydrogen sulphide may be formed by the decomposition of certain metallic sulphides, like that of calcium, in the presence of moisture, as suggested by Anderson and Flett with regard to certain muds at the

Soufriere of St Vincent. Sulphur dioxide, SO_2, is one of the commonest exhalations, especially at acid fumaroles. It may be detected by its characteristic smell, that of burning brimstone, even when present in very small proportion and in the presence of an excess of hydrochloric acid. By hydration it readily forms sulphurous acid, which may be further oxidized to sulphuric acid. J. B. Boussingault found free sulphuric acid (with hydrochloric acid) in the water of the Rio Vinagre which issues from the volcano of Purace in the Andes of Colombia; and it occurs also in certain other volcanic waters. Carbon dioxide, CO_2, is generally a product of the later stages of an eruption and is often evolved after all other gases have ceased to escape.

Although it may sometimes be due to the decomposition of limestone, it seems to be mostly of true magmatic origin. At the well-known Grotta del Cane, at Lake Agnano, in the Phlegraean Fields near Naples, there has been for ages a copious discharge, and analyses of the air of the cave by T. Graham Young showed the presence of from 61.5 to 7 1% of carbon dioxide.

Gautier, in 1907, found 96 to 97% of this gas in the vapours (excluding water-vapour) emitted from the Solfatara near Pozzuoli in the Bay of Naples. The gas by its density tends to accumulate in depressed areas, as in the Death Gulch in the Yellowstone Park and in the Upas Valley of Java. In the Eifel, in the Auvergne and in many other volcanic regions it is discharged at temperatures not above that of the atmosphere.

This natural carbonic acid gas is now utilized industrially at many localities. In the gases of the fumaroles of Mont Pele, carbon monoxide, CO, was detected by H. Moissan. Probably certain hydrocarbons, notably methane or marsh-gas, CH_4, often exist in volcanic gases.

They might be formed by the action of water on natural carbides, such as that of magnesium, calcium, &c. Moissan found 5.46% of methane in vapour from a fumarole on Mont Pele in 1902. Free hydrogen was detected by R. Bunsen as far back as 1846 in vapours from volcanoes in Iceland.

In 1861 Deville and Fouque found it, with hydrocarbons, at Torre del Greco near Naples; and in 1866 Fouque discovered it at Santorin, where some of the vapour at the immediate focus of eruption contained as much as 30% of hydrogen. It is notable that at Santorin free oxygen was also found.

The elements of water may possibly exist, at the high temperature of the magma, in a state of dissociation, and certain volcanic explosions have sometimes been attributed to the combination of these elements.

Oxygen is not infrequently found among volcanic emanations, but may perhaps be derived in most cases from superficial air and ground-water; and in like manner the nitrogen, often detected, may be sometimes of atmospheric origin, though in other cases derived from nitrides in the lava. In the vapours emitted by Mont Pele in 1902 argon was detected by H. Moissan, to the extent of 0.71%; and in those from Vesuvius in 1906 argon and neon were found by Gautier.

The collection of volcanic vapours offers difficulty, and it is not easy to avoid admixture with the atmosphere. F. A. Perret has successfully collected gases on,Vesuvius.

VOLCANIC FLAMES

Although the incandescence of the lava and stones projected during an eruption, and the reflection from incandescent matter in the crater have often been mistaken for red flames, there can be no doubt that true combustion, though generally feeble, does occur during volcanic outbursts.

Among the gases cited above, hydrogen, hydrogen sulphide and the hydrocarbons are inflammable. The flames seen in volcanoes are generally pale and of bluish, greenish or yellowish tint. The presence of hydrogen, sodium and hydrocarbons in the volcanic flames of Kilauea. These, however, were referred to the kindling of sulphur deposited around the fumaroles, the flames being coloured by the presence of boric acid and arsenic sulphide.

When a stream of lava flows over vegetation the combustion of the leaves and wood may be mistaken for flames issuing from the lava. In like manner brushwood may grow in the crater of a dormant volcano and be ignited by a fresh outburst of lava, thus producing flames which, from their position in the crater, may readily deceive an observer.

Volcanic Sublimates

Certain mineral substances occur as sublimates in and around the volcanic vents, forming incrustations on the lava. They are either deposited directly from the effluent vapours, which carry them in a volatile condition, or are produced by interaction of the vapours among themselves; whilst some of the incrustations, rather loosely called sublimates, are due to reaction of the vapours on the constituents of the lava.

Possibly at the temperature of the magma-reservoirs even silica and various silicates may be volatilized, and might thus yield sublimation products. Many of the volcanic sublimates occur at first as incandescent crusts on the lava. Being generally unstable they are difficult of preservation, and are not usually well represented in collections. Among the commonest sublimates is halite, or sodium chloride, $NaCl$, occurring as a white crystalline incrustation, sometimes accompanied, as at Vesuvius, by sylvite, or potassium chloride, KCl, which forms a similar sublimate. The two chlorides may be intimately associated.

Sal ammoniac, or ammonium chloride, $NH_4 Cl$, is not uncommon, especially at Etna, as a white crystalline crust, probably formed in part by the reaction of hydrochloric acid with nitrogen and hydrogen in the vapours. Bunsen, on finding it in Iceland, regarded it as a product of the distillation of organic matter.

At the Solfatara, near Pozzuoli, sal ammoniac was formerly collected as a sublimate on tiles placed round a bocca or vapourvent. Ferric chloride, $FeCl_3$, not infrequently occurs as a reddish or brownish yellow deliquescent incrustation, and because it thus colours the lava it has received the name of molysite.

The action of hydrochloric acid on the iron compounds in the lava may readily yield this chloride, which from its yellowish colour has sometimes been mistaken for sulphur. A crystalline sublimate from the fumaroles on Vesuvius, containing ferric and alkaline chlorides,

$$KCl\ NH_4\ Cl\ _2FeCl\ _3 + 6H_2O,$$

is known as kremersite, after P. Kremers. From a scoriaceous lava found on Vesuvius after the eruption of 1906, Johnston-Lavis procured a yellow rhombohedral sublimate, which he proved to be a chloride of manganese and potassium, whence he proposed for it the name chlormanganokalite.

It was studied by L. J. Spencer, and found to contain $4KCl.MnCl_2$. Chlorocalcite, or native calcium chloride, $CaCl_2$, has been found in cubic crystals on Vesuvian lava. Fluorite, or calcium fluoride, CaF_2, is also known as a volcanic product. Lead chloride, $PbCl_2$, a rare Vesuvian mineral, was named cotunnite, after Dr Cotugno of Naples.

The action of hydrogen sulphide on this chloride may give rise to galena, PbS, found by A. Lacroix on Vesuvius in 1906. Atacamite, or cupric oxychloride, $CuCl_2.3Cu\ (OH)_2$, occurs as a green incrustation on certain Vesuvian lavas, notably those of 1631. Another green mineral from Vesuvius was found by A. Scacchi to be a sulphate containing copper, with potassium and sodium, which he named from its fine colour *euclorina* - a word which has been written in English as euchlorinite.

The copper in the sublimates on Vesuvius will sometimes plate the iron nails of a traveller's boots when crossing the newly erupted lava. Cupric oxide, CuO, occurs in delicate crystalline scales termed tenorite, after Professor G. Tenore of Naples; whilst cupric sulphide, CuS, forms a delicately reticulated incrustation known as covellite, after N. Covelli, its discoverer at Vesuvius.

A sublimate not infrequently found in feathery crystalline deposits on lava at Vesuvius, and formerly called "Vesuvian salt," is a potassium and sodium sulphate, $(K. Na)_2 SO_4$, known as aphthitalite (from Gr. a4B%Tos, imperishable, and &As, salt).

A sulphate with the composition $PbSO_4 (K Na)_2 SO_4$, found in the fumaroles at Vesuvius after the eruption of 1906, was named by A. Lacroix palmierite, after L. Palmieri, who was formerly director of the observatory on Vesuvius. Various sulphites are formed on lavas by the sulphurous acid of the vapours.

Ferric oxide, Fe_2O_3, which occurs in beautiful metallic scales as specular iron-ore, or as an amorphous reddish incrustation on the lava, is probably formed in most cases by the interaction of vapour of ferric chloride and steam at a high temperature.

Less frequently, magnetite, Fe_3O_4, and magnesioferrite, $MgFe_2O_4$, are found in octahedral crystals on lava. An iron nitride (Fe_5N_2) was detected thinly incrusting a lava erupted at Etna in 1874, and was named by O. Silvestri, who examined it, siderazote.

Boric acid, H_3BO_3, occurs in the crater of Vulcano so abundantly that it was at one time collected commercially. It has also led to the foundation of an industry in Tuscany, where it is obtained from the soffioni of the Maremma.

From Sasso in Tuscany it has received the name of sassolin or sassolite. Realgar, or arsenic sulphide, As_2S_2, occurs in certain volcanic exhalations and is deposited as an orange-red incrustation, often associated with sulphur, as at the Solfatara, where orpiment, As_2S_3, has also been found.

Of all volcanic products, sulphur is in some respects the most important. It may, in large quantity lining the walls of the crater, as at Popocatepetl in Mexico, where it was formerly worked by the Indian "volcaneros," or on the other hand it may be a rare product, as at Vesuvius.

Sulphur appears generally to owe its origin in volcanic areas to the interaction of sulphur dioxide and sulphuretted hydrogen, or to the action of water on the latter. A volcanic vent where sulphur is deposited is truly a solfatara (*solfo terra*) or a soufriere, but all volcanoes which have passed into that stage in which they emit merely heated vapours now pass under this name. The famous Solfatara, an old crater in the Phlegraean Fields, exhales sulphurous vapours, especially at the Bocca Grande, from which sulphur is deposited. In the orangecoloured sulphur of the Solfatara, realgar may be present to the extent of as much as 18%. A brown seleniferous sulphur occurring at Vulcano, one of the Lipari Islands, was termed by W. Haidinger volcanite, but it should be noted that Professor W. H. Hobbs has applied this name to an anorthoclase-augite rock ejected as bombs at Vulcano. Sulphur containing selenium is known as a volcanic product in Hawaii, whilst in Japan not only selenium but tellurium occurs in certain kinds of sulphur.

At the Solfatara, near Pozzuoli, the hot sulphurous vapours attack the trachytic rocks from which they issue, giving rise to such products as alum, kaolin and gypsum. To some of these products, including alunogen and mendozite (soda-alum), the name solfatarite was given by C. W. Sheppard in 1835.

By prolonged action of the acid vapours on lava, the bases of the silicates may be removed, leaving the silica as a soft white chalk-like substance. The occurrence of kaolin and other white earthy alteration-products led to the hills around the Solfatara being known to the Romans as the *Colli leucogei.*

THE HOT DUST CLOUD AND AVALANCHE OF PELE

The terrific eruptions in the islands of Martinique and St Vincent in the West Indies in 1902, furnished examples of a type of activity not previously recognized by vulcanologists, though, as Professor A. Lacroix has pointed out, similar phenomena have no doubt occurred elsewhere, especially in the Azores.

By Drs Tempest Anderson and J. S. Flett, who were commissioned by the Royal Society to report on the phenomena, this type of explosive eruption is distinguished as the " Pelean type." Its distinctive character is found in the sudden emission of a dense black cloud of superheated and suffocating gases, heavily charged with incandescent dust, moving with great velocity and accompanied by the discharge of immense volumes of volcanic sand, which are not rained down in the normal manner, but descend like a hot avalanche.

The cloud, with the avalanche, is called by Lacroix a *nuee Peleenne,* or *nuee ardente,* the latter term having been applied to the fatal cloud in the eruptions at San Jorge in the Azores in 1818. In its typical form, the cloud seen at Pele appeared as a solid bank, opaque and impenetrable, but having the edge in places hanging like folds of a curtain, and apparently of brown or purplish colour. Rolling along like an inky torrent, it produced in its passage intense darkness, relieved by vivid lightning. So much solid matter was suspended in the cloud, that it became too dense to surmount obstacles and behaved rather like a liquid. It has, however, been suggested that its peculiar movement as it swept down the mountain was due not simply to its heavy charge of solids, but partly to the oblique direction of the initial explosion. After leaving the crater, it underwent enormous expansion, and Anderson and Flett were led to suggest that possibly at the moment of emission it might have been partly in the form of liquid drops, which on solidifying evolved large volumes of gas held previously in occlusion.

The deadly effect of the blast seems to have been mostly due to the irritation of the mucous membrane of the respiratory passages by the fine hot dust, but suffocating gases, like sulphur dioxide and sulphuretted hydrogen, were associated with the water-vapour. Possibly the incandescent dust was even hotter than the surrounding vapour, since the latter might be cooled by expansion.

It is said that the black cloud as it swept along was accompanied by an indraught of air, not however sufficiently powerful to check its rapid advance. The current of air was likened by Anderson and Flett to the inrush of air at a railway station as an express train passes. An attempt was made to determine the temperature of the fatal blast which destroyed St Pierre, but without very definite results.

Thus it was assumed that as the telephone wires were not melted the temperature was below the fusing-point of copper: possibly, however, the blast may have passed too rapidly to produce the effects which might normally be due to its temperature.

SHAPE OF VOLCANIC CONES

Those volcanic products which are solid when ejected, or which solidify after extrusion, tend to form by their accumulation around the eruptive vent a hill, which, though generally more or less conical, is subject to much variation in shape. It occasionally happens that the hill is composed wholly of ejected blocks, not themselves of volcanic origin.

In this case an explosion has rent the ground, and the effluent vapours have hurled forth fragments of the shattered rock through which the vent was opened, but no ash or other fragmentary volcanic material has been ejected, nor has any lava been poured forth. This exceptional type is represented in the Eifel by certain monticules which consist mainly of fragments of Devonian slate, more or less altered.

In some cases the area within a ring of such rocky materials is occupied by a sheet of water, forming a crater-lake, known in the Eifel as a *rnaar*. Piles of fragmentary matter of this character, though containing neither cinders nor lava, may be fairly regarded as volcanic, inasmuch as they are due to the explosive action of hot subterranean vapours. In the ordinary paroxysmal type of eruption, however, cinders and ashes are shot upwards by the explosion and then descend in showers, forming around the orifice a mound, in shape rather like the diminutive cone of sand in the lower lobe of an hour-glass.

Little cindercones of this character may be formed within the crater of a large volcano during a single eruption; whilst large cones are built up by many successive discharges, each sheet of fragmentary material mantling more or less regularly round the preceding layer.

The symmetry of the hill is not infrequently affected by disturbing influences - a strong wind, for example, blowing the loose matter towards one side. The sides of a cinder cone have generally a steep slope, varying from 30 0 to 45°, depending on the angle of repose of the ejectamenta.

Excellent examples of small scoria-cones are found among the puys of Auvergne in central France, whilst a mag nificent illustration of this type of hill is furnished by Fusiyama, in Japan, which reaches an altitude of 12,000 ft. How such a cone may be rapidly built up was well shown by the formation of Monte Nuovo, near Pozzuoli - a hill 400 ft. high and a mile and a half in circumference, which is known from contemporary evidence to have been formed in the course of a few days in September 1538.

The shape of a cinder cone may be retained for ages, since it is not liable to suffer greatly by denudation, as the rain soaks into the loose porous mass instead of running down the outside. If lava rises in the duct of a cinder cone, it may, on accumulation in the crater, break down the wall, and thus effect its escape as a stream. Cones breached in this way are not uncommon in Auvergne.

It often happens that the cinders and ashes ejected from a volcano become mixed with water, and so form a paste, which sets readily as a hard tufaceous mass. Such natural tuff is indeed similar to the hydraulic cement known as pozzolana, which is formed artificially from volcanic ashes, and is renowned for durability.

Although streams of volcanic mud are commonly associated with the ashes of a cinder-cone they may also form independent structures or tuffcones. These are generally broad-topped hills, having sides with an angle of slope as low in some cases as 15°.

Lava-cones are built up of streams of lava which have consolidated around the funnel of escape. Associated with the lava, however, there is usually more or less fragmentary matter, so that the cones are composite in structure and consequently more acute in shape than if they were composed wholly of lava.

As the streams of lava in a volcano run at different times in different directions, they radiate from the centre, or flow from lateral or eccentric orifices, as irregular tongues, and do not generally form continuous sheets covering the mountain.

When lava is the sole or chief element in the cone, the shape of the hill is determined to a great extent by the chemical composition and viscosity of the lava, its copiousness and the rapidity of flow. If the lava be highly basic and very mobile, it may spread to a great distance before solidifying, and thus form a hill covering a large area and rising perhaps to a great height, but remarkably flat in profile. Were the lava perfectly liquid, it would indeed form a sheet without any perceptible slope of surface.

As a matter of fact, some lavas are so fluent as to run down an incline of 1°, and flat cones of basalt have in some cases a slope of only 10° or even less. The colossal mass of Mauna Loa, in Hawaii, forms a remarkably flat broad cone, spreading over a base of enormous area and rising to a height of 13,900 ft. Major Dutton, writing in 1883, said that "a; moderate eruption of Mauna Loa represents more material than Vesuvius has emitted since the days of Pompeii."

Yet the lava is so mobile that it generally wells forth quietly, without explosive demonstration, and therefore unaccompanied by fragmentary ejectamenta. Fluent lavas like those of Hawaii are also poured forth from the volcanoes and volcanic fissures of Iceland.

If the lava be less basic and less fusible, the hill formed by its accumulation instead of being a low dome will take the shape of a cone with sides of higher gradient: in the case of andesite cones, for instance, the slope may vary from 25° to 35°. Acid rocks, or those rich in silica, such as rhyolites and trachytes, may be emitted as very viscous lavas tending to form dome-shaped or bulbous masses.

Experiment shows that such lavas may persist for a considerable time in a semi-solid condition. It is possible for a dome to increase in size not by the lava running over the crater and down the sides but by injection of the pasty magma within the expanding bulb while still soft; or if solidified, the crust yields by cracking.

Such a mode of growth, in which the dome consists of successive sheets that have been compared to the skins of an onion, has been illustrated by the experiments of Dr A. Reyer, and the structure is typically represented by the mamelons or steep-sided domes of the Isle of Bourbon.

The Puy-de-Dome in Auvergne is an example of a cone formed of the trachytic rock called from its locality domite, whilst the Grand Sarcoui in the same region illustrates the broad domeshaped type of hill. Such domes may have no summit-crater, and it is then usually assumed that the top with the crater has been removed by denudation, but possibly in some cases such a feature never existed.

The " dome volcano " of von Seebach is a dome of acid lava extruded as a homogeneous mass, without conspicuous chimney or crater. Although domes are usually composed of acid rocks, it seems possible that they may be formed also of basic lavas, if the magma be protruded slowly at a low temperature so as to be rapidly congealed.

The Spine of Pele

A peculiar volcanic structure appeared at Mont Pele in the course of the eruption of 1902, and was the subject of careful study by Professor A. Lacroix, Dr E. A. Hoovey, A. Heilprin and other observers. It appears that from fissures in the floor of the Etang Sec a viscous andesitic lava, partly quartziferous, was poured forth and rapidly solidified superficially, forming a domeshaped mass invested by a crust or carapace. According to Lacroix, the crust soon became fractured, partly by shrinkage on consolidation and partly by internal tension, and the dome grew rapidly by injection of molten matter.

Then there gradually rose from the dome a huge monolith or needle, forming a terminal spine, which in the course of its existence varied in shape and height, having been at its maximum in July 1903, when its absolute height was about 5276 ft. above sea-level. The walls of the spine, inclined at from 75° to 90° to the horizon, were apparently slickensided, or polished and scratched by friction: masses were occasionally detached and vapours were continually escaping. Several smaller needles were also formed.

Some observers regarded the great spine as a solidified plug of lava from a previous outburst, expelled on a renewal of activity. Lacroix, however, believed that it was formed by the extrusion of an enormous mass of highly viscid magma, perhaps partly solidified before emission, and he compared the formation of the dome in the crater to the structure on Santorin in 1866, described by Fouque as a " cumulo-volcano." Professor H. F. Cleland has suggested a comparison with the cone of andesite in the crater of the volcano of Toluca in Mexico, and it is said that similar formations have been observed in the volcanoes of the Andes. Dr Tempest Anderson, on visiting Pele in 1907, found a stump of the spine, consisting of a kind of volcanic agglomerate, rising from a cone of talus formed of its ruins.

The Crater

The eruptive orifice in normal volcano - the *bocca* of Italian vulcanologists - is usually situated at the bottom of a depression or cup, known as the crater. This hollow is formed and kept open by the explosive force of the elastic vapours, and when the volcano becomes dormant or extinct it may be closed, partly by rock falling from its crumbling walls and partly by the solidification of the lava which it may contain.

If a renewed outburst occurs, the floor of the old crater may reopen or a new outlet may be formed at some weak point on the side of the mountain: hence a crater may, with regard to position, be either terminal or lateral. The position of the crater will evidently be also changed on any shifting of the general axis of eruption. In shape and size the crater varies from time to time, the walls being perhaps breached or even blown away during an outburst.

Hence the height of a volcanic mountain in activity, measured to the rim of the crater or the terminal peak, is not constant. Vesuvius, for example, suffered a reduction of several hundred feet during the great eruption of 1906, the east side of the cone having lost, according to V. R. Matteucci, 120 metres.

Whilst in many cases the crater is a comparatively small circular hollow around the orifice of discharge, it forms in others a large bowllike cavity, such as is termed in some localities a " caldera." In the Sandwich Islands the craters are wide pits bounded by nearly vertical walls, showing stratified and terraced lavas and floored by a great plain of black basalt, sometimes with lakes of molten lava.

Professor W. H. Pickering compares the lava-pits of Hawaii to the crater-rings in the moon. Some of the pit-craters in the Sandwich Islands are of great size, but none comparable with the greatest of the lunar craters. Dr G. K. Gilbert, however, has suggested that the ring-shaped pits on the moon are not of volcanic origin, but are depressions formed by the impact of meteorites.

Similarly the "crater" of Coon Butte, near Canyon Diablo, in Arizona, which is 4000 ft. in diameter and 500 ft. deep, has been regarded as a vast pit due to collision of a meteorite of prodigious size.

Probably the largest terrestrial volcanic crater is that of Aso-san, in the isle of Kiushiu, which is a huge oval depression estimated by some observers to have an area of at least zoo sq. m. Some of the large pit-craters have probably been formed by subsidence, the cone of a volcano having been eviscerated by extravasation of lava, and the roof of the cavity having then subsided by loss of support. The term caldera has sometimes been limited to craters formed by such collapse.

On the floor of the crater, ejected matter may accumulate as a conoidal pile; and if such action be repeated in the crater of the new cone, a succession of concentric cones will ultimately be formed. The walls of a perfect crater form a ring, giving the cone a truncated appearance, but the ring may suffer more or less destruction in the course of the history of the mountain.

A familiar instance of such change is afforded by Vesuvius. The mountain now so called, using the term in a restricted sense, is a huge composite cone built up within an old crateral hollow, the walls of which still rise as an encircling rampart on the N. and N.E. sides, and are known as Monte Somma; but the S. and S.W. sides of the ancient crater have disappeared, having been blown away during some former outburst, probably the Plinian eruption of 79.

In like manner the relics of an old crater form an amphitheatre partially engirdling the Soufriere in St Vincent, and other examples of " Somma rings " are known to vulcanologists.

Much of the fragmental matter ejected from a volcano rolls down the inside of the crater, forming beds of tuff which incline towards the central axis, or have a centroclinal dip. On the contrary, the sheets of cinder and lava which form the bulk of the cone slope away from the axis, or have a dip that is sometimes described as pericentric or qua-qua-versal.

According to the old "crater-of-elevation theory," held especially by A. von Humboldt, L. von Buch and Elie de Beaumont, this inclination of the beds was regarded as mainly due to upheaval. It was contended that the volcanic cone owed its shape, for the most part, to local distension of the ground, and was indeed comparable to a huge blister of the earth's crust, burst at the summit to form the " elevation crater." Palma, in the Canary Islands, was cited as a typical example of such a formation.

This view was opposed mainly by Poulett-Scrope, Sir Charles Lyell and Constant Prevost, who argued that the volcano, so far from being bladder-like, was practically a solid cone of erupted matter: hence this view came to be known as the " crater-of-eruption theory." Its general soundness has been demonstfated whenever an insight has been obtained into the internal structure of a volcano.

Thus, after the eruption of Krakatoa in 1883 a magnificent natural section of the great cone of Rakata, at the S. end of the island, was exposed - the northern half having been blown away - and it was then evident that this mountain was practically a solid cone, built up of a great succession of irregular beds of tuff and lava, braced together by intersecting dykes. The internal architecture of a volcano is rarely so well displayed as in this case, but dissections of cones, more or less distinct, are often obtained by denudation.

In connexion with the structures called *laccoliths,* there may have been an elevation, or folding, and even faulting, of the superficial rocks by subterranean intrusion of lava; but this is different from the local expansion and rupture of the ground required by the old theory. It may be noted, however, that in recent years the view of elevation, in a modified form, has not been without supporters.

Where the growth of a volcanic mound takes place from within, as in certain steep-sided trachytic cones, there may be no perceptible crater or external outlet. Again, there are many volcanoes which have no crater at the summit, because the eruptions always take place from lateral outlets. Even when a terminal pit is present, the lava may issue from the body of the mountain, and in some cases it exudes from so many vents or cracks that the volcano has been described as " sweating fire."

PARASITIC CONES

In the case of a lofty volcano the column of lava may not have sufficient ascensional force to reach the crater at the summit, or at any rate it finds easier means of egress at some weak spot, often along radial cracks, on the flanks of the mountain.

Thus at Etna, which rises to a height of more than io,800 ft., the eruptions usually proceed from lateral fissures, sometimes at least half-way down the mountain-side. When fragmental materials are ejected from a lateral vent a cinder-cone is formed, and by frequent repetition of such ejections the flanks of Etna have become dotted over with hundreds of scoria-cones much like the puys of Auvergne, the largest (Monte Minardo) rising to a height of as much as 750 ft. Hills of this character, seated on the parent mountain, are known as parasitic cones, minor cones, lateral cones.

Such subordinate cones often show a tendency to a linear arrangement, rising from vents or *bocche* along the floor of a line of fissure. Thus in 1892 a chain of five cones arose from a rift on the S. side of Etna, running in a N. and S. direction, and the hills became known as the Monti Silvestri, after Professor Orazio Silvestri of Catania. This rift, however,was but a continuation of a fissure from which there arose in 1886 the series of cones called the Monti Gemmellaro, while this in turn was a prolongation of a rent opened in 1883.

The eruption on Etna in the spring of 1910 took place along the same general direction, but at a much higher elevation. The tendency for eruptions to be renewed along old lines of weakness, which can be readily opened afresh and extended, is a feature well known to vulcanologists.

The small cones which are frequently thrown up on lava streams were admirably exemplified on Vesuvius in the eruption of 1855 and figured by J. Schmidt. The name of "driblet cones" was given by J. D. Dana to the little cones and pillars formed by jets of lava projected from blowing holes at Kilauea, the drops of lava remaining plastic and cohering as they fell.

Such clots may form columns and. pyramids, with almost vertical sides. Steep-sided cones more or less of this character occur elsewhere, but are usually built up around spiracles. Small cones formed by mere dabs of lava are known trivially as "spatter cones."

FISSURE ERUPTIONS

In certain parts of the world there are vast tracts of basaltic lava with little or no evidence of cones or of pyroclastic accompaniment. To explain their formation, Baron F. von Richthofen suggested that they represent great floods of lava which were poured forth not from ordinary volcanic craters with more or less explosive violence, but from great fissures in the earth's crust, whence they may have quietly welled forth and spread as a deluge over the surface of the country. The eruptions were thus effusive rather than explosive.

Such phenomena, constituting a distinct type of vulcanism, are distinguished as fissure eruptions or massive eruptions - terms which suggest the mode of extrusion and the character of the extruded matter.

As the lava in such outflows must be very fusible, it is generally of basaltic type, like that of Hawaii: indeed, the Hawaiian volcanoes, with their quiet emission of highly fluent lavas, connect the fissure eruptions with the " central eruptions," which are usually regarded as representing the normal type of activity.

At the present day true fissure eruptions seem to be of rather limited occurrence, but excellent examples are furnished by Iceland. Here there are vast fields of black basalt, formed of sheets of lava which have issued from long chasms, studded in most cases with rows of small cones, but these generally so insignificant that they make no scenic features and might be readily obliterated by denudation. Dr T. Thoroddsen enumerates 87 great rifts and lines of cones in Iceland, and even the larger cones of Vesuvian type are situated on fissures.

It is believed that fissure eruptions must have played a far more important part in the history of the earth than eruptions of the familiar cone-and-crater type, the latter representing indeed only a declining phase of vulcanism. Sir Archibald Geikie, who has specially studied the subject of fissure eruptions, regards the Tertiary basaltic plateaus of N.E. Ireland and the Inner Hebrides as outflows from fissures, which may be represented by the gigantic system of dykes that form so marked a feature in the geological structure of the northern part of Britain and Ireland.

These dykes extend over an area of something like 40,000 sq. m., while the outflows form an aggregate of about 3000 ft. in thickness. In parts of Nevada, Idaho, Oregon and Washington, sheets of late Tertiary basalt from fissure eruptions occupy an area of about 200,000 sq. m., and constitute a pile at least 2000 ft. thick. In India the "Deccan traps" represent enormous masses of volcanic matter, probably of like origin but of Cretaceous date, whilst South Africa furnishes other examples of similar outflows.

Professor J. W. Gregory recognized in the Kapte plains of East Africa evidence of a type of vulcanism, which he distinguished as that of "plateau eruptions." According to him a number of vents opened at the points of intersection of lines of weakness in a high plateau, giving rise to many small cones, and the simultaneous flows of lava from these cones united to form a far-spreading sheet.

Extrusive and Intrusive Magmas

When the molten magma in the interior of the earth makes its way upwards and flows forth superficially as a stream of lava, the product is described as extrusive, effusive, effluent or eruptive; but if, failing to reach the surface, the magma solidifies in a fissure or other subterranean cavity, it is said to be intrusive or irruptive.

Rocks of the former group only are sometimes recognized as strictly "volcanic," but the term is conveniently extended, at least in certain cases, to igneous rocks of the latter type, including therefore certain hypabyssal and even plutonic rocks.

When the intrusive magma has been forced into narrow irregular crevices it forms "veins," which may exhibit complex ramifications, especially marked in some acid rocks; but when injected into a regularly shaped fissure, more or less parallel-sided, and cutting across the planes of bedding, it forms a wall-like mass of rock termed a "dyke." Most dykes are approximately vertical, or at least highly inclined in position.

The inclination of a dyke to a vertical plane is termed its "hade." In a cindercone, the lava as it rises may force its way into cracks, formed by pressure of the magma and tension of the vapours, and will thus form a system of veins and dykes, often radiating from the volcanic axis and strengthening the structure by binding the loose materials together.

Thus, in the Valle del Bove, a huge cavity on the east side of Etna, the walls exhibit numerous vertical dykes, which by their hardness stand out as rocky ribs, forming a marked feature in the scenery of the valley.

In a similar way dykes traverse the walls of the old crater of Monte Somma at Vesuvius. Exceptionally a dyke may be hollow, the lava having solidified as a crust at the margin of the fissure but having escaped from the interior while still liquid.

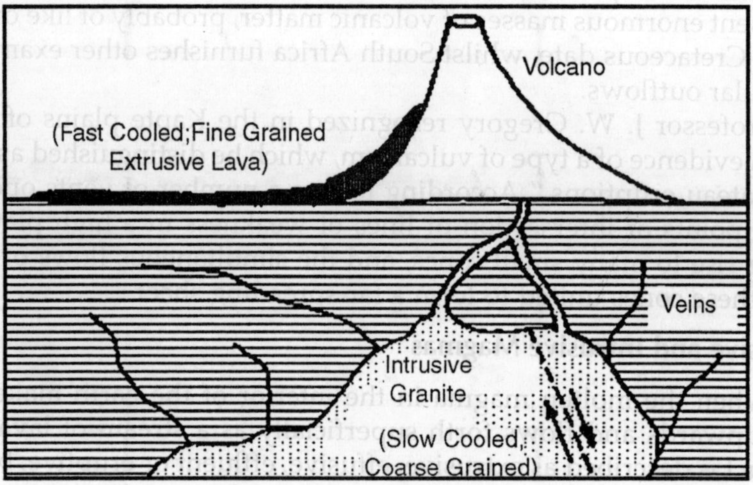

Fig. Extrusive and Intrusive Magmas

When molten matter is thrust between beds of tuff or between successive lava-flows or even ordinary sedimentary strata, it forms an intrusive sheet of volcanic rock known as a " sill. " A sill may sometimes be traced to its connexion with a dyke, which represents the channel up which the lava rose, but instead of reaching the surface the fluid found an easier path between the strata or perhaps along a horizontal rent.

Although a dyke may represent a conduit for the ascent of lava which has flowed out superficially, yet if the lava has been removed at the surface by denudation the dyke terminates abruptly, so that its function as the former feeder of a lava-current is not evident. In other cases a dyke may end bluntly because the crack which it occupies never reached the surface.

Lava which has insinuated itself between planes of stratification may, instead of spreading out as a sheet or sill, accumulate locally as a lenticular mass, known as a *laccolith* or *laccolite.* Such a mass, in many cases rather mushroom-shaped, may force the superincumbent rocks upwards as a dome, and though at first concealed may be ultimately exposed by removal of the overlying burden by erosion. The term *phacolite* was introduced by A. Harker to denote a meniscus-shaped mass of lava intruded in folded strata, along a crest or a trough.

The *bysmalith* of Professor Iddings is a *laccolith* of rather plug-like shape, with a faulted roof. An intrusive mass quite irregular in shape has been termed by R. A. Daly a *chonolith* whilst an intrusion of very great size and illdefined form is sometimes described as a *bathylith* or *batholite.*

STRUCTURAL PECULIARITIES IN LAVA

Many of the structures exhibited by lava are due to the conditions under which solidification has been effected.

A dyke, for example, may be vitreous at the margin where it has been rapidly chilled by contact with the walls of the fissure into which it was injected, whilst the main body may be lithoidal or crystalline: hence a basalt dyke will sometimes have a selvage formed of the basaltic glass known as tachylyte.

A similar glass may form a thin crust on certain lava-flows. In a homogeneous vitreous lava, contraction on solidification may develop curved fissures, well seen in the delicate " perlitic " cracks of certain obsidians, indicating a tendency to assume a globular structure. This structure becomes very distinct by the development of " spherulites, " or globular masses with a radiating fibrous structure, sometimes well seen in devitrified glass.

Occasionally the spherulitic bodies in lava are hollow, when they are known as lithophyses, of which excellent examples occur at Obsidian Cliff in the Yellowstone National Park, as described by Professor Iddings. Globular structure on a large scale is sometimes displayed by lavas, expecially those of basic type, such as the basalt of Aci Castello in Sicily, which was probably formed, according to Professor Gaetano Platania, by flow of the lava into submarine silt, relics of which still occur between the spheroids.

Joints, or cracks formed by shrinkage on solidification, may divide a sheet of lava into columns, as familiarly seen in basalt, where the rock often consists of a close mass of regular polygonal prisms, mostly hexagonal. Each prism is divided at intervals by transverse joints, more or less curved, so that the portions are united by a slight ball-and-socket articulation. As the long axes of the columns lie at right angles to the cooling surface they are vertical in a horizontal sheet of lava, horizontal in a vertical dyke, and inclined or curved in other cases.

It sometimes happens that in a basaltic dyke the formation of the prisms, having started from the opposite walls as chilling surfaces, has not been completed; and hence the prisms fail to meet in the middle.

A spheroidal structure is often developed in basalt columns by weathering, the rock exfoliating in spherical shells, rather like the skins of an onion: such a structure is characteristically shown at the Kasekellar, known also as the Elfen Grotto, at Bertrich, near Alf on the Mosel, where the pillars of the lava are broken into short segments which suggest by their flattened globular shape a pile of Dutch cheeses. Although prismatic jointing, or columnar structure, is most common in basalt, it occurs also in other volcanic rocks.

Fine columns of obsidian, for instance, are seen at Obsidian Cliff in the Yellowstone Park, where the pillars may be 50 ft. or more in height. Such an occurrence, however, is exceptional.

Vitreous lavas often show fluxion structure in the form of streaks, bands or trains of incipient crystals, indicating the flow of the mass when viscous. The character of this structure is related to the viscosity of the lava.

Those structural peculiarities which depend mainly on the presence of vapour, such as vesiculation, have been already noticed, and the porphyritic structure has likewise been described.

Submarine Volcanoes

Considering how large a proportion of the face of the earth is covered by the sea, it seems likely that 'volcanic eruptions must frequently occur on the ocean-floor. When, as occasionally though not often happens, the effects of a submarine eruption are observed during the disturbance, it is seen that the surface of the sea is violently agitated, with copious discharge of steam; the water passes into a state of ebullition, perhaps throwing up huge fountains; shoals of dead fishes, with volcanic cinders, bombs and fragments of pumice, float around the centre of eruption, and ultimately a little island may appear above sea-level.

This new land is the peak of a volcanic cone which is based on the sea-floor, and if in deep water the submarine mountain must evidently be of great magnitude. Christmas Island in the Indian Ocean, described by Dr C. W. Andrews, appears to be a volcanic mountain, with Tertiary limestones,. standing in water more than 14,000 ft. deep.

Many volcanic islands, such as those abundantly scattered over the Pacific, must have started as submarine volcanoes which reached the surface either by continued upward growth or by upheaval of the sea-bottom. Etna began its long geological history by submarine eruptions in a bay of the Mediterranean, and Vesuvius in like manner represents what was originally a volcano on the sea-floor.

As the ejectamenta from a submarine vent accumulate on the sea-bottom they become intermingled with relics of marine organisms, and thus form fossiliferous volcanic tuffs. By the distribution of the ashes over the sea-floor, through the agency of waves and currents, these tuffs may pass insensibly into submarine deposits of normal sedimentary type.

One of the best examples of a submarine eruption resulting in the formation of a temporary island occurred in 1831 in the Mediterranean between Sicily and the coast of Africa, where the water was known to have previously had a depth of too fathoms.

After the usual manifestations of volcanic activity an accumulation of black cinders and ashes formed an island which reached at one point a height of 200 ft., so that the pile of erupted matter had a thickness of about 800 ft.

The new island, which was studied by Constant Prevost, became known in England as Graham's Island, in France as Ile Julie and in Italy by various names as Isola Ferdinandea. Being merely a loose pile of scoriae, it rapidly suffered erosion by the sea, and in about three months was reduced to a shoal called Graham's Reef.

In 1891 a submarine eruption occurred near the isle of Pantellaria in the same waters, and the eruptive centre was termed by Professor H. S. Washington and Foerstner volcano, but it gave rise to no island.

A well-known instance of a temporary volcanic island was furnished by Sabrina - an islet of cinders thrown up by submarine eruptions in 1811, off the coast of St Michael's, one of the Azores. The island of Bogosloff, or Castle island, in Bering Sea, about 40 m. W. of Unalaska Island, is a volcanic mass which was first observed in 1796 after an eruption.

In 1883 another eruption in the neighbouring water threw up a new volcanic cone of black sand and ashes, known as New Bogosloff or Fire Island, situated about half a mile to the N.W. of Old Bogosloff, with which it was connected by a low beach. Another island, called Perry Island, larger than either of the others, made its appearance in the neighbourhood about the time of the great earthquake in California in 1906. It is reported that some of these islands have since disappeared.

Mud Volcanoes

Mud volcanoes are small conical hills of clay which discharge, more or less persistently, streams of fine mud, sometimes associated with naphtha or petroleum, and usually with bubbles of gas. As the mud is generally saline, the hills are known also as " salses." The gases are chiefly hydrocarbons, often with more or less sulphuretted hydrogen and carbon dioxide, and sometimes with nitrogen. Though generally less than a yard in height, the cones may in exceptional cases rise to an elevation of as much as Soo ft.

The mud oozes from the top and spreads over the sides, or is spurted forth with the gases. Occasionally the discharge is vigorous, mud and stones being thrown up to a considerable height, sometimes accompanied by flames due to combustion of the hydrocarbons.

Mud volcanoes occur in groups, and have a wide distribution. They are known in Iceland; in Modena; at Taman and Kertch, in the Crimea; at Baku on the Caspian; in Java and in Trinidad: Humboldt described those near Turbaco, in Colombia.

In Sicily they occur near Girgenti, and a group is known at Paterno on Etna. Bythe Sicilians they are termed, *maccalube,* a word of Arabic origin. The " paint-pots " of the Yellowstone National Park are small mud volcanoes.

Many, so-called mud volcanoes appear to be due to the derangement of subterranean water-flow or to landslips in connexion with earthquakes, whilst others may be referable to certain chemical reactions going on underground; but there are others again which seem to be truly of volcanic origin.

Hot water and steam escaping through clays, or crumbling tuffs reduced to a clayey condition, may form conical mounds of pasty material, through which mud oozes and water escapes.

Geysers are closely related to volcanoes, but in consequence of their special interest they are treated separately. For natural steam-holes and other phenomena connected with declining vulcanicity.

GEOGRAPHICAL DISTRIBUTION OF VOLCANOES

It is matter of frequent observation that volcanoes are most abundant in regions marked by great seismic activity. Although the volcano and the earthquake are not usually connected in the direct relation of cause and effect, yet in many cases they seem referable to a common origin.

Both volcanic extrusion and crustal movement may be the means of relieving local strains in the earth's crust, and both are found to occur, as might reasonably be expected, in many parts of the earth where folding and fracture of the rocks have frequently happened and where mountain-making appears to be still in progress.

Thus, volcanoes may often be traced along zones of crustal deformation, or folded mountain-chains, especially where they run near the borders of the oceanic basins. They are frequently associated with the Pacific type of coast-line.

The most conspicuous example of linear distribution is furnished by the great belt of volcanoes, coinciding for the most part with a band of seismic disturbance, which engirdles intermittently the huge basin of the Pacific; though here, as elsewhere in studying volcanic topography, regard must be paid to dormant and extinct centres as well as to those that are active at the present time.

As volcanoes are in many cases ranged along what are commonly regarded as lines of fracture, it is not surprising that the centres of most intense vulcanicity are in many cases situated at the intersection of two or more fracture-lines. On the eastern side of the Pacific Ocean the great volcanic ring may be traced, though with many and extensive interruptions, from Cape Horn to Alaska.

In South America the chain of the Andes between Corcovado in the south and Tolima in the north is studded at irregular intervals with volcanoes, some recent and many more extinct, including the loftiest volcanic mountains. in the world.

The linear arrangement, often a marked feature in the distribution of volcanoes, is well exemplified in the general north-andsouth trend of the Andean ranges, the volcanoes being situated along the orographic axis. These folded mountains with their volcanoes also illustrate the close relationship to the sea so frequently observed in volcanic topography, a relationship, however, not without many exceptions.

The volcanic rock called andesite was so named by L. von Buch from its characteristic occurrence in the Andes. It is notable that the volcanic rocks throughout the great Pacific belt present much similarity in composition. The volcanoes of Ecuador have been described in detail by A. Stiibel and others.

Central America contains a large number of active volcanoes and solfataras, many of which are located in the mountains parallel to the western coast. Conseguina, on the south side of the Gulf of Fonseca, is remarkable for its eruption in 1835, when an enormous, volume of ash was ejected and the summit of the mountain blown away.

Izalco, in San Salvador, came into existence in 1770, and is habitually active. In the centre of Lake Ilopango in Salvador,. which possibly occupies an ancient crater, a volcanic island arose in 1880 and attained a height of 160 ft. Guatemala is peculiarly rich in volcanoes, as described by Dr Tempest Anderson, who visited the country in 1907.

The Cerro Quemado, or the Volcano of Quezaltenango, was the scene of a great eruption in 1785. At the Volcano of Santa Maria there was an outburst in 1902 more violent than the simultaneous eruptions in the Lesser Antilles. The cones of Guatemala include the Volcan de Fuego and the Volcan de Agua, the former often active in historic times, whilst the latter is notable for the flood which in 1541 swept down from the mountain and destroyed Old Guatemala, but this flood was probably not of volcanic origin.

The plateau of Mexico is the seat of several active volcanoes which occur in a band stretching across the country from Colima in the west to Tuxtla near Vera Cruz. The highest of these volcanic mountains is Orizaba, or Cithaltepetl, rising to an altitude of 18,200 feet, and known to have been active in the 16th century.

Popocatepetl (" the smoking mountain ") reaches a height of about. 17,880 ft., and from its crater sulphur was at one time systematically collected. It came into existence rapidly during an eruption which began in September 1759, when it was said by unscientific observers that the ground became suddenly inflated from below.

The cone, though not of exceptional magnitude, is situated in an elevated district, and its summit rises to about 4330 ft. above sea-level. In the neighbourhood of Jorullo there are three subordinate cones of similar character known as *volcancitos,* with great numbers of small mounds of cinder and ash formed around fumaroles on the lava, and locally called *hornitos,* or " little ovens. " The streams of basaltic lava from Jorullo form rough barren surfaces, which pass under the name of *malpays,* or bad lands.

In the United States very few volcanoes are active at the present day, though many have become extinct only in times that are geologically recent. An eruption occurred in 1857 at Tres Virgines, in the south of California, and the cinder cone on Lassen's Peak was also active in the middle of the 19th century.

The Mono Valley craters and Mount Shasta, in California, are extinct. The Cascade Range contains numerous volcanic peaks, but only few show signs of activity. 1Vlount Hood, in Oregon, exhales vapour, as also does Mount Rainier in Washington. Mount St Helens was in eruption in 1841 and 1842; and Mount Baker the most northern of the volcanoes connected with the Cascade Range, is said to have been active in 1843.

Few volcanic peaks occur in the Rocky Mountains, but evidence of lingering activity is very marked in the geysers and hot springs of the Yellowstone National Park. The earth's internal heat is also manifested at many points elsewhere, as at Steamboat Springs on the Virginia Range, an offshoot of the Sierra Nevada, and in the Comstock Lode.

Volcanic activity is prominent in Alaska, along the Coast Range and in the neighbouring islands. The crater of Mount Edgecumbe, in Lazarus Island, is said to have been active in 1796, but this is doubtful.

Mount Fairweather has probably been in recent activity, and the lofty cone of Mount Wrangell, on Copper river, is reported to have been in eruption in 1819. In the neighbourhood of Cook's Inlet there are several volcanoes, including the island of St Augustine. Unimak Island has two volcanoes, which have supplied the natives, with sulphur and obsidian; one of these volcanoes being Mount Shishaldin, a cone rivalling Fusiyama in graceful contour. The.

Aleutian volcanic belt is a narrow, curved chain of islands, extending from Cook's Inlet westwards for nearly 1600 m. It is notable that the convexity of the curve faces the great ocean, as has been observed in other cases, the arcs following the direction of the rock-folds. According to Professor I. C. Russell, an authority on the volcanoes of N. America, there are in the Aleutian Islands and in the peninsula no fewer than 57 craters, either active or recently extinct.

From the Aleutian Islands the volcanic band of the Pacific changes its direction, and passing to the peninsula of Kamschatka, where 14 volcanoes are said to be active, turns southwards and forms the festoon of the Kurile Islands. Here again the convexity of the insular arc is directed towards the ocean.

This volcanic archipelago leads on to the great islands of Japan, where the volcanoes have been studied by Professor J. Milne, who also described those of the Kuriles. Of the 54 volcanoes recognized as now active or only recently extinct in Japan, the best known is the graceful cone of the sacred mountain Fusiyama, but others less pretentious are far more dangerous.

The great eruption of Bandaisan, about 120 m. N. of Tokio, which occurred in 1888, blew off one side of the peak called Kobandai, removing, according to Professor Sekiya's estimate, about 2982 million tons of material. Aso-san in Kiushui, the southernmost large island of Japan, is notable for the enormous size of its crater. In the Bonin group of islands volcanic activity is indicated by such names as Volcano Island and Sulphur Island.

South of the Japanese archipelago the train of volcanoes passes through some small islands in or near the Loo Choo group and thence onwards by Formosa to the Philippine Islands, where subterranean activity finds abundant expression in earthquakes and volcanoes.

After leaving this region the linear arrangement of the eruptive centres becomes less distinctly marked, for almost every island in the Moluccas and the Sunda Archipelago teems with volcanoes, solfataras and hot springs.

Possibly, however, a broken zone may be traced from the Moluccas through New Guinea and thence to New Zealand, perhaps through eastern Australia (for though no active volcanoes are known there, relics of comparatively recent activity are abundant); or again by way of the Bismarck Archipelago, the Solomon Islands, the New Hebrides, the Fiji Islands and Kermodoc Island.

The great volcanic district in New Zealand is situated in the northern part of North Island, memorable for the eruption of Tarawera in 1886. This three-peaked mountain on the south side of Lake Tarawera, not previously known to have been active, suddenly burst into action; a huge rift opened, and Lake Rotomahana subsided, with destruction of the famous sinter terraces. The crater of Tongariro is in the solfatara stage, whilst Mount Ruapehu is regarded as extinct. On White Island in the Bay of Plenty the cone of Wharkari is feebly active.

Far to the south, on Ross Island, off South Victoria Land, in Antarctica, are the volcanoes of Erebus and Terror, the former of which is active. These are often regarded as remotely related to the Pacific zone, but Dr G. T. Prior has shown that the Antarctic volcanic rocks which he examined belonged to the Atlantic and not the Pacific type.

Within the great basin of the Pacific, imperfectly surrounded by its broken girdle of volcanoes, there is a vast number of scattered islands and groups of islands of volcanic origin, rising from deep water, and hiving in many cases active craters. The most important group is the Hawaiian Archipelago, where there is a chain of at least fifteen large volcanic mountains - all extinct, however, with the exception of three in Hawaii, namely Mauna Loa, Kilauea and Hualalai; and of these Hualalai has been dormant since 1811.

It is notable that the two present gigantic centres of activity, though within 20 m. of each other, appear to be independent in their eruptivity. Several of the Hawaiian Islands, as pointed out by J. D. Dana, who was a very high authority on this group, consist of two volcanoes united at the base, forming volcanic twins or doublets.

The volcanic regions of the Pacific are connected with those of the Indian Ocean by a grand train of islands rich in volcanoes, stretching from the west of New Guinea through the Moluccas and the Sunda Islands, where they form a band extending axially through Java and Sumatra. Here is situated the principal theatre of terrestrial vulcanicity, apparently representing an enormous fissure, or system of fissures, in the earth's crust, sweeping in a bold curve, with its convexity towards the Indian Ocean.

Numerous volcanic peaks occur in the string of small islands to the east of Java - notably in Flores, Sumbawa, Lombok and Bali; and one of the most terrific eruptions on record in any part of the world occurred in the province of Tomboro, in the island of Sumbawa, in the year 1815. Java contains within its small area as many as 49 great volcanic mountains - active, dormant and extinct.

The largest is Smerin, about 12,000 ft. high, but the most regularly active is said to be Gownong Lamongang, which is in almost uninterrupted activity, emitting usually only ashes and vapour, though in 1883 lava streamed forth. Many of the Javanese volcanoes present marked regularity of contour, with the sides of the cones rather symmetrically furrowed by tropical rains and probably ridged by ash-slides. The radial furrows on volcanic cones are sometimes known as " barrancos."

The little uninhabited island of Krakatoa in the Strait of Sunda appears to be situated at a volcanic node, or the intersection of two curved fissures, and it is believed that the island itself represents part of the basal wreck of what was once a volcano of gigantic size. After two centuries of repose, a violent catastrophe occurred in 1883, whereby the greater part of the island was blown away. This eruption and its effects were made the subject of careful study by Verbeek, B g on and Judd.

Through the great island of Sumatra, a chain of volcanoes runs longitudinally, and may possibly be continued northwards in the Bay of Bengal by Barren Island and Norcondam - the former an active and the latter an extinct volcano. On the western side of the Indian Ocean a small volcanic band may be traced in the islands of the Mascarene group, several craters in Reunion being still active.

Far south in the Indian Ocean are the volcanic islands of New Amsterdam and St Paul. The Comoro Islands in the channel of Mozambique exhibit volcanic activity, whilst in East and Central Africa there are several centres, mostly extinct but some partially active, associated with the Rift Valleys.

The enormous cones of Kenia and Kilimanjaroo are extinct, but on Kibo, one of the summits of the latter, a crater is still preserved. The Mfumbiro volcanoes, S. of Lake Edward, rise to a height of more than 14,700 feet. Kirunga, N. of Lake Kivu, is still partially active. Elgon is an old volcanic peak, but Ruwenzori is not of volcanic origin. On the west side of Africa, the Cameroon Peak is a volcano which was active in 1909, and the island of Fernando Po is also volcanic. Along the Red Sea there are not wanting several examples of volcanoes, such as Jebel Teir. Aden is situated in an old crater.

Passing to the Atlantic, a broken band of volcanoes, recent and extinct, may be traced longitudinally through certain islands, some of which rise from the great submarine ridge that divides the ocean, in part of its length, into an eastern and a western trough. The northern extremity of the series is found in Jan Mayen, an island in the Arctic Ocean, where an eruption occurred in 1818. Iceland, however, with its wealth of volcanoes and geysers, is the most important of all the Atlantic centres.

According to Dr T. Thoroddsen there are in Iceland about 130 post-glacial volcanoes, and it is known that from 25 to 30 have been in eruption during the historic period. Many of the Icelandic lava-flows, such as the immense flood from Laki in 1783, are referable to fissure eruptions, which are the characteristic though not the exclusive form of activity in this island. Probably this type was also responsible for the sheets of old lava in the terraced hills of the Faroe Islands, to which may have been related the Tertiary volcanoes of the west of Scotland and the north of Ireland.

An immense gap separates the old volcanic area of Britain from the volcanic archipelagoes of the Azores, the Canaries and the Cape Verd Islands. Palma - a little island in the Canary group, with a caldera or large crater at its summit, from which fissures or barrancos radiate - is famous in the history of vulcanology, in that it furnished L. von Buch with evidence on which he founded the " crater-ofelevation " theory.

The remaining volcanic islands of the Atlantic chain, all now cold and silent, include Ascension, St Helena and Tristan da Cunha, whilst in the western part of the South Atlantic are the small volcanic isles of Trinidad and Ferdinando do Noronha. St Paul's rocks appear also to be of volcanic origin.

One of the most important volcanic regions of the world is found in the West Indies, where the Lesser Antilles - the scene of the great catastrophes of 1902 - form a string of islands, stretching in a regular arc that sweeps in a N. and S. direction across the eastern end of the Caribbean Sea. Subject to frequent seismic disturbance, and rich in volcanoes, solfataras and hot springs, these islands seem to form the summit of a great earth-fold which, rising as a curved ridge from deep water, separates the Caribbean Sea from the Atlantic.

The volcanoes are situated on the inner border of the curve. It is notable that the Antilles and the Sunda Islands, two of the grandest theatres of vulcanicity on the face of the earth, are situated at the antipodes of each other - one being apparently an eastern and the other a western offshoot of the great Pacific girdle.

The European volcanoes, recent and extinct, may be regarded as representing rather ill-defined branches thrown off eastwards from the Atlantic band. Vesuvius is the only active volcano on the mainland, but in the Mediterranean there are Etna on the coast of Sicily; the Lipari

Islands, with Stromboli and Vulcano in chronic activity; and farther to the east the archipelago of Santorin, where new islands have appeared in historic times. Submarine eruptions have occurred also between Sicily and the coast of Africa; one in 1831 having given rise temporarily to Graham's Island, and another in 1891 appearing near Pantellaria, itself a volcanic isle.

Of the extinct European volcanoes, some of the best known are in Auvergne, in the Eifel, in Bohemia and in Catalonia, whilst the volcanic land of Italy includes the Euganean hills, the Alban hills, the Phlegraean Fields, &c. The great lakes of Bolsena and Bracciano occupy old craters, and many smaller sheets of water are on similar sites.

The volcanic islands no longer active include Ischia, with the great cone of Epomeo which was in a state of eruption in 1301; the Ponza Islands, Nisida, Vivera and others near Naples; and several in the Greek archipelago, such as Milos, Kimolos and Polinos.

From the eastern end of the Mediterranean evidence of former volcanic activity may be traced into Asia Minor and thence to Armenia and the Caucasus. East of Smyrna there is a great desolate tract which the ancients recognized as volcanic and termed the Catacecaumene.

The volcanic districts of Lydia were studied by Professor H. S. Washington. In the plateau of Armenia there are several extinct volcanic mountains, more or less destroyed, of which the best known is Ararat. Nimrud Dagh on the shore of Lake Van is said to have been in eruption in the year 1441. Dr F. Oswald has described the volcanoes of Armenia. Of the volcanoes in Persian territory not now active, Demavend, south of the Caspian, is an important example. Elburz is also described as an old volcano. It has been said that in Central Asia there are certain vents still active, and recent volcanic rocks are known from the Przhevalsky chain and other localities.

The number of volcanoes known to be actually active on the earth is generally estimated at between 300 and 400, but there is reason to believe that this estimate is far too low. If account be taken of those volcanic cones which have not been active in historic time, the total will probably rise to several thousands.

The distribution of volcanoes at various periods of the earth's history, as revealed by the local occurrence of volcanic rocks at different horizons in the crust of the earth. Periods of great earthmovement have been marked by exceptional volcanic activity.

Causes of Vulcanicity

The cause of vulcanicity two problems demand attention: first the origin of the heat necessary for the manifestation of volcanic phenomena, and secondly the nature of the force by which the heated matter is raised to the surface and ejected. According to the old view, which assumed that the earth was a spheroid of molten matter invested by a comparatively thin crust of solid rock, the explanation of the phenomena appeared fairly simple.

The molten interior supplied the heated matter, while the shrinkage of the cooling crust produced fractures that formed the volcanic channels through which it was assumed the magma might be squeezed out in the process of contraction. When physicists urged the necessity of assuming that the globe was practically solid, vulcanologists were constrained to modify their views.

Following a suggestion of W. Hopkins of Cambridge, they supposed that the magma, instead of existing in a general central cavity, was located in comparatively small subterranean lakes. Some authorities again, like the Rev. O. Fisher, regarded the magma as constituting a liquid zone, intermediate between a solid core and a solid shell.

If solidification of the primitive molten globe proceeded from the centre outwards, so as to form a sphere practically solid, it is conceivable that portions of the original magma might nevertheless be retained in cavities, and thus form "residual lakes." Although the mass might be for the most part solid, the outer portion, or "crust," could conceivably have a honeycombed structure, and any magma retained in the cells might serve indirectly to feed the volcanoes.

Neighbouring volcanoes seem in some cases to draw their supply of lava from independent sources, favouring the idea of local cisterns or "intercrustal reservoirs." It is probable, however, that subterranean reservoirs of magma, if they exist, do not represent relics of an original fluid condition of the earth, but the molten material may be merely rock which has become fused locally by a temporary development of heat or more likely, by a relief of pressure.

It should be noted that the quantity of magma required to supply the most copious lava-flows is comparatively small, the greatest recorded outflow ot having exceeded, it is said, six cubic miles; and even this estimate is probably too high. Whilst in many cases the magma-cisterns may be comparatively small and temporary, it must be remembered that there are regions where the volcanic rocks are so similar throughout as to suggest a common origin, thus needing intercrustal reservoirs of great extent and capacity.

It has been suggested that comparatively small basins, feeding individual volcanoes, may draw their supply from more extensive reservoirs at greater depths.

Much speculation has been rife as to the source of the heat required for the local melting of rock. Chemical action has naturally been suggested, especially that of superficial water, but its adequacy may be doubted.

After Sir Humphry Davy's dis covery of the metals of the alkalis, he thought that their remarkable behaviour with water might explain the origin of subterranean heat; and in more recent years others have seen a local source of heat in the oxidation of large deposits of iron, such as that brought up in the basalt of Disco Island in Greenland.

It has been assumed by Moissan and by Gautier that water might attack certain metallic carbides, if they occur as subterranean deposits, and give rise to some of the products characteristic of volcanoes. But it seems that all such action must be very limited, and utterly inadequate to the general explanation of volcanic phenomena.

At the same time it must be remembered that access of water to a rock already heated may have an important. physical effect by reducing its melting point, and may thus greatly assist in the production of a supply of molten matter. The admission of surface-waters to heated rocks is naturally regarded as an important source of motive power in consequence of the sudden generation of vapour, but it is doubtful to what extent it may contribute, if at all, to the origin of volcanic heat.

According to Robert Mallet a competent source of subterranean heat for volcanic phenomena might be derived from the transformation of the mechanical work of compressing and crushing parts of the crust of the earth as a consequence of secular contraction. This view he worked out with much ingenuity, supporting it by mathematical reasoning and an appeal to experimental evidence. It was claimed for the theory that it explained the linear distribution of volcanoes, their relation to mountain chains, the shallow depth of the foci and the intermittence of eruptive activity.

A grave objection,. however, is the difficulty of conceiving that the heat, whether due to crushing or compression, could be concentrated locally so as to produce a sufficient elevation of temperature for melting the rocks. According to the calculations of Rev. O. Fisher, the crushing could not, under the most favourable circumstances, evolve heat enough to account for volcanic phenomena.

Since pressure raises the melting-point of any solid that expands on liquefaction, it has been conjectured that many deep-seated rocks, though actually solid, may be potentially liquid; that is, they are maintained in a solid state by pressure: only. Any local relief of pressure, such as might occur in the folding and faulting of rocks, would tend, without further accession of heat, to induce fusion. But although moderate pressure raises the fusing-point of most solids, it is believed,, from modern researches, that very great pressures may have a contrary effect.

It is held by Professor S. Arrhenius that at great depths in. the earth the molten rock, being above its critical point, can. exist only in the gaseous condition; but a gas under enormous pressure may behave, so far as compressibility is concerned, like a rigid solid.

He concludes, from the high density of the earth as a whole and from other considerations, that the central part of our planet consists of gaseous iron (about 80% of the earth's diameter) followed by a zone of rock magma in a gaseous condition (about 15%), which passes insensibly outwards into liquid rock (4%), covered by a thin solid crust (less than i % of diameter).

If water from the crust penetrates by osmosis through the sea-floor to the molten interior, it acts, at the high temperature, as an acid, and decomposes the silicates of the magma. The liquid rock, expanded and rendered more mobile by this water, rises in fissures, but in its ascent suffers cooling, so that the water then loses its power as an acid and is displaced by silicic acid, when the escaping steam gives rise to the explosive phenomena of the volcano.

The earth was formed by the accretion of vast numbers of small cosmical bodies called planetesimals, the original heat of the earth's interior was due chiefly to the compression of the growing globe by its own gravity. The heat, proceeding from the centre outwards, caused local fusion of the rocks, though without forming distinct reservoirs of molten magma, and the fused matter charged with gases rose in liquid threads or tongues, which worked their way upwards, some reaching the superficial part of the earth and escaping through fissures in the zone of fracture, thus giving rise to volcanic phenomena.

It is held that the explosive activity of a volcano is due to the presence of gases which have been brought up from the interior of the earth, whilst only a small and perhaps insignificant part is played by water of superficial origin.

Entirely new views of the origin of the earth's internal heat have resulted from the discovery of radioactivity. An ingenious hypothesis was enunciated by Major C. E. Dutton, who found in the radioactivity of the rocks a sufficient source of heat for the explanation of all volcanic phenomena.

He believes that the development of heat arising from radioactivity may gradually bring about the local melting of the rocks so as to form large subterranean pools of magma, from which the volcanoes may be supplied. The supply is usually drawn from shallow sources, probably, according to Dutton, from a depth of not more than three or rarely four miles, and in some cases at not more than a mile from the surface.

If the water in the local magma should attain sufficient expansive power, it will rupture the overlying rocks and thus give rise to a volcanic eruption. When the reservoir becomes exhausted the eruption ceases, but if more heat be generated by continued radioactivity further fusion may ensue, and in time the eruption be repeated. According, however, to Professor Joly, it is improbable that sufficient heat for the manifestation of volcanic phenomena could be developed by the local radioactivity of the rocks in the upper part of the earth's crust.

Index